# 投影学习：理论及应用

刘本永　著

科学出版社
北　京

## 内 容 简 介

本书对机器学习中的投影学习算法进行系统的理论分析与应用探讨，内容包括描述型投影学习和鉴别型投影学习之正投影及斜投影学习准则、一般形式、增量学习和稀疏化算法，以及这些算法在信号分析处理、模式识别等领域中的典型应用等。

本书可供人工智能领域有关科研人员参考，也可作为高等院校人工智能、计算机科学与技术、电子信息工程、通信工程、电气工程及其自动化等专业高年级本科生和相应学科研究生的教学参考书，还可供在各种应用场景下从事人工智能技术开发的工程人员参考。

#### 图书在版编目（CIP）数据

投影学习：理论及应用 / 刘本永著. -- 北京：科学出版社，2025.3.
ISBN 978-7-03-081282-7

Ⅰ. TP181

中国国家版本馆 CIP 数据核字第 2025TN7889 号

责任编辑：叶苏苏　贺江艳 / 责任校对：彭　映
责任印制：罗　科 / 封面设计：陈　敬

科学出版社 出版
北京东黄城根北街 16 号
邮政编码：100717
http://www.sciencep.com
四川煤田地质制图印务有限责任公司 印刷
科学出版社发行　各地新华书店经销

\*

2025 年 3 月第 一 版　开本：720×1000　1/16
2025 年 3 月第一次印刷　印张：9 1/2
字数：192 000
**定价：139.00 元**
（如有印装质量问题，我社负责调换）

# 前　言

机器学习是让机器从应用场景的典型事例中总结经验教训、形成认知规律、提升解决相应问题和适应场景变化之能力的过程，其内容包括模型选择（认知规律的表示空间选择）、学习算法（优化准则及求解算法）和设计实现（算法的软硬件设计和实现）等，是实现机器智能/人工智能的基本途径。

本书作为人工智能领域的学术性专著，重点对机器学习中投影学习的基本理论、算法实现和典型应用进行分析和探讨。全书共 5 章，包括：

第 1 章，在概述人工智能与机器学习的基础上，重点论述投影学习的背景、现状和主要问题。

第 2 章，主要论述投影学习的代数与泛函基础，包括希尔伯特空间及再生核希尔伯特空间中的线性算子理论、框架理论和线性算子广义逆理论，等等。

第 3 章，主要探讨描述型投影学习，包括学习的基本准则及其基本形式、扩展形式和增量形式。

第 4 章，重点讨论鉴别型投影学习，包括核非线性鉴别子、表示型核非线性鉴别子、斜投影核鉴别子的基本形式和自适应训练（增量形式和稀疏化方法）。

第 5 章，从信号分析处理和模式识别两个领域介绍投影学习理论和算法的典型应用，包括基于直方图拟合与分解的图像分割、基于曲面拟合与再采样的图像放大、基于多帧融合的图像超分辨重建、基于曲线拟合的语音端点检测与增强，以及手写数字识别、人脸识别、说话人识别、雷达目标识别、视频目标行为识别等应用。

作为实现人工智能的基本途径，机器学习长期成为国内外相关科技工作者的一个优选研究方向和主要研究课题。“九五”期间，作者以雷达目标成像识别为应用场景，承担了预研基金课题“集群目标成像识别研究”（基金号：JD70972），对这一方向中有关投影学习理论和算法进行了初步探索。以此为基础，作者于“十五”初期在日本东京工业大学小川英光实验室就日本文部省基盘 B 项目“神经网络正则化理论及应用研究”从事博士后研究，以及于“十五”后期主持完成教育部科技研究重点项目“目标一维像序列稀疏参数建模理论及应用研究”（任务号：105150），在以上工作中对投影学习进一步展开了系统研究。

长期以来，尤其是“十一五”以来，国家有关部委在人工智能领域积极跟进和持续发力，不断加强相关理论研究和技术开发工作的支持力度。这一时期，作者组织的项目组先后承担了国家自然科学基金课题“增量型核回归分析理论及应用研究”(任务号：60862003)和教育部高等学校博士学科点专项科研基金课题“增量型斜投影学习理论及应用研究”(博导类，任务号：20095201110002)，对已有投影学习理论和算法展开了深入的研究和拓展。同一时期，以作者为团队负责人的项目组承担了科技部国际合作司的重点专项项目“公共安全中数字视听资料分析检验关键技术研究及系统开发”(任务号：2009DFR10530)，为上述基金课题的理论和算法成果提供了具有现实意义的应用场景和应用尝试。

本书主要取材于上述科研课题和技术研发项目的理论算法和应用成果。这些成果凝聚了项目组成员的心血，这些成员包括日本东京工业大学小川英光教授及其研究室人员，贵州省公安厅物证鉴定中心原主任廖翔警官及唐畅、徐德志、胡力航等警官，贵州大学计算机科学与技术学院教师刘洪副教授、梁建娟讲师、徐晶讲师，以及作者先后招收的大部分硕士研究生和博士研究生。在此一并表示感谢！

在相关材料的组织中，作者对绝大部分内容已进行过反复推敲。但是，由于投影学习理论和算法本身对数学基础的要求极高，且实际应用场景往往难以满足理论和算法所需的必要条件，加之时间仓促，所以难免有疏漏和不足之处，衷心希望得到读者的批评指正。

# 目　　录

# 第1章 概　　论

## 1.1 引　　言

本书的主题隶属于人工智能领域的机器学习方向。为此，本章在对人工智能基本概念、发展简史、主要研究领域等进行简要介绍的基础上，对机器学习涉及的基本问题和典型算法进行概述，并进一步对正则化学习、最优泛化投影学习理论和算法进行综述。

## 1.2 人工智能与机器学习概述

本节主要对人工智能的基本概念、发展简史、主要研究领域、关键基础问题和机器学习的基本概念、基本问题、典型算法进行概述。

### 1.2.1 人工智能的基本概念及发展简史

人工智能（artificial intelligence，AI）又叫机器智能（machine intelligence），是指由人工赋予机器的类似于人进行思考、推理和行动等功能的智能。

人工智能是紧随二战后期数字电子计算机（后来简称计算机）的诞生而产生的，其研制的主要目的是快速处理复杂的计算任务。在处理逻辑问题上，计算神经科学创始人麦卡洛克和数学家皮茨于 1943 年提出，通过模仿人类大脑中神经元的协同工作方式建立人工神经元模型，并以适当的结构和连接方式形成人工神经网络（后来简称神经网络），实现基本逻辑运算（与、或、非等）和逻辑推理；并且，他们指出，设计合适的网络具有学习能力（如后来的赫布学习、强化学习）[1-3]。同时，在计算机上模拟人的智能以形成自动机这一问题，也于 1948 年被作为计算机之父的冯·诺依曼提出并展开了奠基性工作[4]。随后，“机器能否思考”这一问题，由作为人工智能之父的图灵于 1950 年在其开山之作《计算之机器与智能》中提出（该文同时还提出一种用于判定机器是否具有智能的试验方法，后来被称为“图灵测试”）[5]。此后有关机器智能的研究工作便如雨后春笋一般，在计算机科学、数学、控制论、信息论、心理学、语言学、神经科学、哲学等多个领域迅速铺开[6-15]。

“人工智能”这一术语，是在 1956 年夏季于美国达特茅斯学院举行的有关机器智能的学术研讨会之准备期（即 1955 年 8 月），由会议发起人麦卡锡、明思基、罗切斯特和香农在其洛克菲勒基金申报书中首次引入的[8, 9]。这次历时近两个月且有计算机科学、数学、心理学、神经科学、信息论等不同学科领域的学者和工程师参加的专题讨论会，同时也标志着人工智能的诞生，因此 1956 年被普遍视为人工智能元年。

关于人工智能的定义有很多说法，包括“行为似人的智能”[6]、“计算智能”[9]、“处理复杂信息问题似人的智能”[11]、“(获取和运用知识）行动精明似人的智能”[12]等，但时至今日，尚无一个能被广泛接受的定义，而本小节一开始的叙述正是综合多种定义的功能性描述[13-16]。尽管如此，自其诞生开始，人工智能的有关学术研究和应用开发就得到不断发展和进步。

从发展历程上看，人工智能主要经历早期、跌宕起伏期、平稳和蓬勃发展期三个阶段。

1. 人工智能的早期

人工智能早期主要指其诞生后近二十年的一段时期。这是人工智能发展的一个黄金时期，其研究开发和应用工作大体从以下三个角度展开：

（1）从认知心理学和信息处理角度，基于逻辑、规则和搜索算法研究让计算机模拟人的思维过程和开展智力活动的方法[10, 17-20]，使计算机具备人类智力活动中诸如数学定理证明和符号积分[21-24]、弈棋[25-29]、翻译[30-33]、问答与会话[34-38]等典型的内在思维和外部行为能力。这是这一时期的主流工作。

（2）从神经心理学和脑功能角度（延续麦卡洛克和皮茨的工作），通过模仿和设计神经系统信息处理的多层结构和神经元之接收、处理和传输信息的功能特点，构造感知机以模拟智慧生物对环境信息（主要是视听觉信息）的感知、处理、识别和理解能力[1-3, 39-51]。

（3）从行为心理学和控制论角度，研究让计算机（或计算机控制的机器人）模拟人的智能行为[52-58]。

事实上，以上三个角度的工作后来逐渐分流并形成了具有明显差异的三大学派，即注重思维过程模拟和逻辑推理的符号主义学派、注重脑功能模拟和神经元联系的连接主义学派，以及注重智慧生物感知和控制行为的行为主义学派，但在人工智能早期三者非常紧密、毫无区分地交织在一起。

这一时期 AI 的核心研究课题包括符号逻辑、自然语言处理与理解、机器感知、语音理解、视觉与图像理解、模式识别、神经网络、机器学习、机器人学等。另外，AI 编程语言的开发在人工智能的发展中也起着举足轻重的作用[17, 59-62]。这一时期的典型成果及其所涉及的核心问题见表 1.1。

**表 1.1　人工智能早期的典型成果及其涉及的主要核心问题**

| 时间 | 成果内容 | 主要核心问题 |
|---|---|---|
| 1950 年 | 第一台神经网络计算机 SNARC（stochastic neural analogy reinforcement computer）诞生（用于在不停尝试中解决问题）[3] | 神经网络，机器学习 |
| 1956 年 | 启发式程序“逻辑理论家”（logic theorist，LT）诞生，证明了世界名著《数学原理》中的 38 个定理[17, 23] | 符号逻辑（搜索推理） |
| | 第一个字符识别程序（the memory test computer，MTC）诞生[40, 41] | 符号逻辑，模式识别 |
| 1957 年 | 通用问题求解机（general problem-solving，GPS）出现[21, 23] | 符号逻辑 |
| 1958 年 | 感知机诞生[42-49] | 神经网络，机器学习 |
| 1959 年 | 西洋跳棋程序在 IBM701 上实现，随后在 IBM7090 上战胜尼雷大师[28] | 符号逻辑，人机交互 |
| | “王浩机”（在 IBM704 上实现）只用 9 分钟就证明了《数学原理》中一阶逻辑的全部定理[63-66] | 符号逻辑 |
| 1960 年 | AI 语言 LISP（list processor）诞生（用于符号处理中罗列指令、操纵逻辑演绎）[59-61] | AI 编程语言 |
| 1961 年 | 第一个模仿人联想记忆和口语学习的模型 EPAM（elementary processors of association memory）诞生[38] | 自然语言处理（词性分析） |
| 1965 年 | 第一个机器翻译系统 SYSTRAN（system analysis translator）诞生[33] | 自然语言处理（语义及上下文分析） |
| | 第一个成功投入使用的专家系统（用于分析化合物的分子结构）“德纳尔（Dendral）”诞生[67-69] | 符号逻辑，知识工程 |
| 1966 年 | 人机对话软件“艾丽莎（Eliza）”（模拟心理治疗专家）诞生[37] | 自然语言处理 |
| | 第一个自主移动机器人“沙基（Shakey）”诞生[70-72] | 自主推理，规划控制 |
| 1968 年 | 语义网络知识表示法诞生[73] | 自然语言处理 |
| 1971 年 | 规划语言 STRIPS（Stanford Research Institute problem solver）诞生[74] | 符号逻辑 |
| 1972 年 | 人机对话系统（用自然语言指挥机器人搭积木动作）“沙尔德鲁（Shrdlu）”诞生[75, 76] | 自然语言理解、启发式搜索、决策与控制 |
| | 人工智能程序设计语言 Prolog（programming in logic）诞生[77] | AI 编程语言 |
| 1973 年 | 第一台双足行走机器人“瓦伯特（Wabot）-1”诞生[78] | 机器视听觉，控制理论与工程，人机交互 |

在这一时期，正是上述这些显著进展、典型成果和研究人员对 AI 未来的乐观预测，让人们对 AI 满怀憧憬和期待，相关研发工作也因此先后得到主要经济体的政府和相关机构（如英国政府、美国国家科学委员会、美国高等研究计划局）的大力支持和持续资助。

2. *人工智能的跌宕起伏期*

这一时期从 20 世纪 70 年代中后期开始，持续十余年。可以分为面对现实的低谷期、短暂的复兴与低迷期和交互式 AI 的曙光期。

1）面对现实的低谷期

这一时期之初，研发人员接续早期工作，在数学定理证明、机器翻译、机器弈棋、机器人研制、专家系统开发等涉及的符号逻辑与启发式搜索算法、自然语言理解、机器学习、机器视听觉、神经网络、知识工程理论与技术等核心问题上持续发力，进一步取得显著的研究成果[79-98]。其中最具轰动效应的理论成果是利用计算机解决了数学界的四色定理证明难题[98]，而针对不同场景并综合有关成果开发的典型应用系统包括“听说（Hearsay）”（用于接收和理解口述语句并在合理时间内做出响应的人机交互系统）、“迈森（MYCIN）”（用于帮助医生诊断传染性血液病的医用专家系统）等，其中集成了自然语言理解、启发式搜索推理、知识工程和知识库、似然推理等核心技术[94-97]。

尽管如此，鉴于以下主要原因，在持续五六年的时间里，AI 处于饱受批评、资助撤回、大量 AI 研究人员被迫调整自己研究方向的低谷时期。

首先，计算机内存、算力等硬件资源非常有限，软件开发也基本上是基于规则编程完成（这是作为主流的符号主义之典型特征），在其中需要明确指定程序行为的所有规则。这些局限必然导致 AI 程序往往只能解决一些规则非常有限、涉及信息量和知识量不大、应用场景非常具体明确的简单问题。在稍微复杂的问题上，消除歧义、正确理解上下文等难题往往限制了 AI 的理解和推理能力，乃至难以达到三岁儿童的智力水平。

其次，并非所有智能问题都涉及逻辑推理，如对模式识别问题，以感知机为象征的连接主义方法更具优势[42-48]。然而，当时的感知机学习算法由于受算力等因素影响而无法推广到多层网络结构，因而无法解决非线性分类、异或运算等问题，乃至作为人工智能先驱的明斯基也因此提出质疑并大加批判[99]，致使众多研发人员产生了不少悲观情绪。

再次，行为主义侧重于应用技术（如机器人）的开发，尚未在某个重要理论方面获得突破后迎来快速增长。

最后，20 世纪 70 年代，全球处于经济低迷、通货膨胀持续、石油危机加深、政治危机突出的年代。

尽管如此，在这一低谷期，依旧有少数研究人员（尤其是欧洲和日本的一些学者）热衷于诸如联想记忆网络（associative memory network，AMN）、自组织网络（self-organizing network，SON）、循环神经网络（recurrent neural network，RNN）、认知机等神经网络和知识工程方面的研究[83-96]，为连接主义的复兴和专家系统的应用打下了坚实的基础。

2）短暂的复苏与低迷期

进入 20 世纪 80 年代，人工智能进入短暂（持续六七年时间）的复苏阶段，主要体现在连接主义的突破性进展和符号主义的现实性应用两个方面。

一方面，学者们在延续低谷期有关多层神经网络学习机制的研究工作上取得了突破，其中误差反向传播（back propagation，BP）算法使神经网络可以自动调节神经元连接强度进而实现不断优化目标函数的学习功能，从理论上彻底解决了多层神经网络训练问题，使有效训练大规模神经网络成为可能，并初步用于玻尔兹曼机（随机 RNN）、卷积神经网络（convolutional neural network，CNN）（改进的认知机）等网络的训练[100-105]。BP 算法为神经网络的快速发展和连接主义的兴盛奠定了坚实的基础，同时标志着人工智能正式进入数据驱动时代。

另一方面，延续低谷期的工作，符号主义在知识工程理论与技术方面取得不断进步并日趋成熟，为开发实用的专家系统奠定了基础并创造了商机[94, 106-113]。这些系统，除了早期仅针对问题可分解的应用而设计的数据解释型专家系统（如“德纳尔”[67-69]、“听说”[95, 96]）和故障诊断型专家系统（如“迈森”[97]），设计型专家系统（如超大规模集成电路设计系统[109]和计算机系统配置设置系统[110]）、规划型专家系统（如生产规划系统[111]）、预测型专家系统（如暴雨预报系统[112]）等各种类型的专家系统纷纷出现，并在工程、生产、医疗、服务等众多行业得到应用和推广。

在这一复苏背景下，主要经济体政府竞相投入 AI 研发计划：日本政府斥巨资支持第五代计算机计划、美国政府成立工业界联盟 MCC（Microelectronics and Computer Consortium，即微电子与计算机联盟）并启动自动陆地车辆（autonomous land vehicle，ALV）计划、英国政府启动耗资巨大的阿尔维计划、欧洲也启动了“欧洲信息技术战略计划”，这些计划都努力把 AI 作为未来经济发展的战略支撑点。但是，这些计划执行了几年时间就困难重重、难以达到预期目标，而得到较为广泛应用的专家系统相较于通用计算机系统而言也普遍存在购置和维护费用高、难以及时升级以适应变化的应用需要等困境，所以相关研发计划到 20 世纪 80 年代后期就陆续中止，使得短暂复苏的 AI 再次进入低迷期。

在此期间，传统 AI 的局限性也进一步被暴露出来：利用预先编写和存储的特定场景知识并通过搜索式推理来实现智能，无法解决应用中普遍面临的场景不确定性、知识不完全性及由此涉及的现场交互问题和可扩放问题[114]。

3）交互式 AI 的曙光期

以机器人为载体的行为主义，强调通过载体与环境的交互（包括感知、处理、决策、反馈等）来增强载体的智能，在交互性和可扩放性方面比传统基于符号推理的 AI 更具有先天优势，为现实地解决 AI 应用问题提供了全新的途径，被称为新人工智能（nouvelle AI）或现场人工智能（situated AI）[114-120]。同时，作为连接主义长期的研究课题，神经网络在环境交互方面有独到之处。因此，神经网络与现场 AI 的集成，为以机器学习（尤其是融交互与反馈为一体的强化学习[3, 121-124]）和神经网络为主体之 AI 的进一步平稳和蓬勃发展带来了曙光。

### 3. 人工智能的平稳和蓬勃发展期

进入 20 世纪 90 年代后，以交互式 AI 为主要特征的人工智能得到了近二十年的平稳发展，在机器学习、神经网络、智能机器人与智能体及其交融方面的研发工作不断取得新成果[124-135]。与此同时，计算机硬件水平、处理速度的巨大提升和通信技术的快速发展，加快了全球数字化和信息化的进程，使得日益成熟和普及的互联网技术、分布式与并行处理技术、实时处理技术[如通用图形处理单元（general-purpose graphical processing units，GP-GPU）]成为让以机器学习和神经网络为主体的交互式 AI 技术落地的助推器。

这一时期，虽然 AI 水平依旧与初期人们期待的智能水平（与人类可比拟的水平，即强 AI 水平）相距甚远，但在解决特定领域的特定问题（这样的 AI 被称为弱 AI）方面，其水平已得到显著提升，实现了一些早期的研发目标：1995 年，华莱士开发的聊天机器人“爱丽丝（Alice）”能够利用互联网不断增进自身数据集并优化内容[136]；1997 年，IBM 开发的象棋电脑“深蓝”（其处理速度是早期弈棋机的 1000 万倍，能在 1s 内计算 2 亿种可能的位置，可搜索并估计随后的 12 步棋）击败了国际象棋冠军卡斯帕罗夫[137, 138]；1998 年和 1999 年，美国太格电子公司（Tiger Electronics）和日本索尼公司先后发明的宠物机器人“菲比（Furby）”和“爱博（Aibo）”，具备视听觉感知和交互学习能力[139, 140]；2000 年，本田公司发布的机器人产品“阿西莫（ASIMO）”，具有听觉感知、交互学习、决策和行动能力[141]；2002 年，美国先进机器人技术公司 iRobot 研制的扫地机器人“扫把（Roomba）”，具备环境监测、记忆学习、决策和行动能力，面向市场获得成功[142]；2005 年和 2007 年，美国斯坦福大学和卡内基梅隆大学的自动驾驶机器人先后赢得美国国防部高级研究计划局设置的技术指标（自动驾驶千米数）和场景（常规行驶环境、一般交规限制条件等）挑战，而谷歌研发的无人驾驶汽车到 2010 年就已经创下 16 万千米无事故的纪录[143-146]；2011 年，苹果公司发布的语音助手“Siri”能根据人的指令和喜好从网络上搜索信息、辅助导航、播报天气等[147]，而 IBM 开发的人工智能程序“沃特森（Watson）”在智力问答节目中战胜了两位人类冠军[148, 149]，等等。

这一时期，深度神经网络及基于 GP-GPU 的大规模无监督学习算法逐渐成熟[150-160]、互联网广泛普及与云计算技术快速进步[161-166]、大规模图像库不断更新和丰富[167-171]，从理论、算法、技术、算力、数据等方面为人工智能的蓬勃发展奠定了坚实基础。

在最近十余年来的蓬勃发展中，人工智能以深度学习为核心[172-174]，以算法、算力和数据为基本要素，在图像识别[175-178]、智能语音交互[179-182]、弈棋[183-186]、蛋白质结构解析和肿瘤分析[187-191]、计算机编程和科学研究等众多问题中大放异彩[192-200]，成为新一轮科技革命和产业变革的强力推进器。

### 1.2.2 人工智能的主要研究领域及关键基础问题

综上所述，传统人工智能是知识驱动型第一代人工智能，而最近十余年蓬勃发展的人工智能是以数据、算力、算法为基本要素的数据驱动型第二代人工智能[201]。从层次上看，第二代人工智能自下而上可分为基础设施层、感知-认知-决策层和应用层这三个层次，各层有关主要科学问题[200]如下：

（1）基础设施层，包括数据的获取、数据的存储与计算、机器学习算法，以及三者融合实现的架构。

（2）感知-认知-决策层，包括视听觉感知、自然语言和知识图谱认知、自动规划与决策等。

（3）应用层，包括在基础科学研究、信息空间交互、社会治理、工业生产、日常生活等方面的应用。

因此，人工智能的研究领域包括知识表示、自动推理、自然语言处理、视听觉感知与处理、智能机器人，等等。

机器学习是支撑这些领域的关键基础，包括回归分析[202]、决策树算法、k-均值（k-means）算法、k-近邻（k-nearest neighbors，k-NN）算法[203]、支持向量机（support vector machine，SVM）算法[125, 130]、随机森林算法等基本学习算法。

神经网络则是机器学习的基本实现形式，包括多层感知机（multi-layer perceptron，MLP）[87, 131]、SON 和自组织映射网络（self-organizing map network，SOMN）[87, 91, 126]、SVM[130]、AMN[83, 86, 89, 90]、玻尔兹曼机[103]、径向基函数神经网络（radial basis function neural network，RBF-NN）[204]等等。

深度学习则是深度 CNN[102, 105, 154]、受限玻尔兹曼机（restricted Boltzmann machine，RBM）[160]、深度自动编码机[150, 155]、生成式对抗网络（generative adversarial network，GAN）[205, 206]、深度置信网络（deep belief nets，DBN）[151, 157, 158, 160]等多层网络（深度网络）的基本训练方式。

总之，人工智能与机器学习、神经网络、深度学习的关系如图 1.1[188, 200, 207]所示。

### 1.2.3 机器学习的基本概念及典型算法

本小节在介绍机器学习基本概念和基本问题的基础上，对机器学习的基本类型和典型算法进行概述。

#### 1. 机器学习的基本概念及基本问题

早期，机器学习（machine learning，ML）泛指无须严格编程就能使机器具有

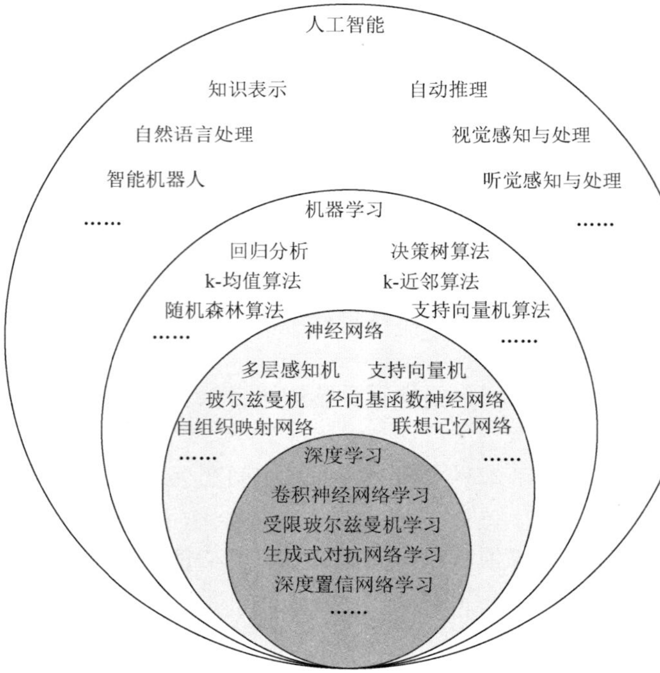


图 1.1 人工智能与机器学习、神经网络、深度学习的基本关系[188, 200, 207]

学习能力的科学领域[28]。在以数据驱动为特征的第二代人工智能中，机器学习作为人工智能和数据科学的核心，一般指机器为完成特定任务、从历史经验或教训中积累知识规律或寻找数据依赖关系并不断提升性能的过程[125, 208-210]。

机器学习的任务可以是随机总体的概率密度估计、函数逼近、机器翻译、模式特征提取、模式分类等数学和人工智能领域的传统任务，也可以是图像目标（如其中的文本）检测、信号噪声抑制、数据压缩、信息编码等信号分析处理任务，还可以是日常知识传授中的学习任务，等等。对应地，潜在的知识规律或依赖关系可以是概率密度函数、未知函数、不同语言的语义关系、模式分类器、知识点，等等。机器学习性能则指机器完成任务的好坏程度，通常用性能指标刻画。

对潜在的知识规律或数据依赖关系可用未知函数 $f_0(\boldsymbol{x})$ 表示的一般情况，机器学习问题可以抽象为如图 1.2 所示的典型系统。图中训练数据集 $D$ 是历史经验或教训用实例具体化的结果，它是按未知的确定规律 $f_0(\boldsymbol{x})$（如联合概率密度函数）产生的实例集合：$D=\{\boldsymbol{x}(n)\}$（$n=1, 2, \cdots, N$），其中 $\boldsymbol{x}(n)=[x_1(n), x_2(n), \cdots, x_M(n)]^{\mathrm{T}}$（T 表示向量或矩阵的转置）是第 $n$ 个实例（$M$ 维向量），$N$ 为实例数即训练样本数。机器学习就是基于训练数据集 $D$，按特定任务和性能要求设计恰当的学习算法得到学习

结果 $f(\boldsymbol{x})$的过程，而学习性能就体现为 $f(\boldsymbol{x})$对 $f_0(\boldsymbol{x})$的逼近程度。因此，数据驱动的机器学习常归结为函数逼近问题[202, 204, 209-212]。

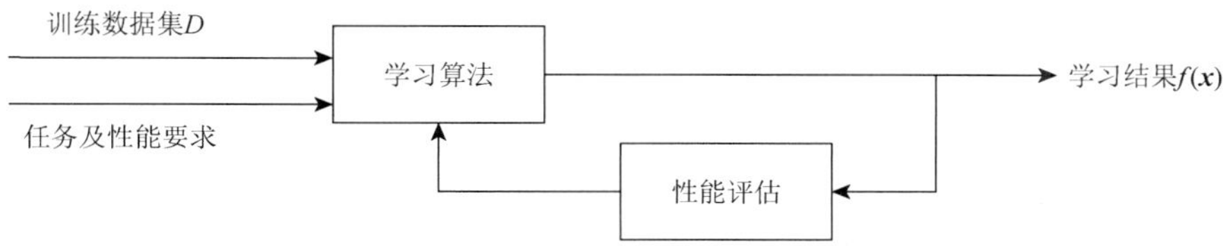


图 1.2 机器学习问题的抽象系统[208, 211]

根据上述基本概念，可以看出机器学习包括模型选择（知识规律的表示）、优化准则及其求解（学习算法设计）和算法实现（硬件结构或软件架构）三个基本问题[212]，而优化准则及其求解的投影法，是本书讨论的重点内容。

1）模型选择问题

模型选择问题是统计学、信息论、控制论、机器学习、信号分析处理等领域的一个共同课题[213-222]。这里的模型，可以是变量关系分析中 $f_0(\boldsymbol{x})$的具体模型（如 SVM、时间序列模型、回归模型、神经网络模型）之规模（如阶数、变量数、隐层数和神经元数）[125, 130, 211-219, 222]，也可以是函数逼近中$f_0(\boldsymbol{x})$所属的函数空间[220-222]。机器学习中的模型选择，就是在所有可能的模型中确定一个恰当的模型，使得学习结果$f(\boldsymbol{x})$对训练样本及训练样本以外实例都有好的性能表现。

对函数空间选择（或信号表示问题）而言，模型选择的示意图如图 1.3[211]所示（一维情况即 $M=1$）：对于图 1.3（a）所示的学习结果 $f(x)$，既可以像图 1.3（b）那样用多个余弦函数表示，也可以像图 1.3（c）那样用多个高斯函数表示，还可以用其他形式的多个函数（即函数系）表示，不同的函数系对应于不同的模型，这时函数系的选择就是模型选择。

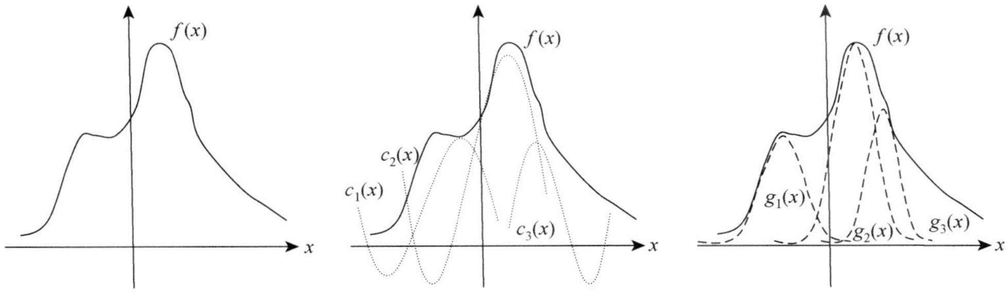


(a) 学习结果 (b) 余弦函数系模型 (c) 高斯函数系模型

图 1.3 模型选择[211]

由于复杂的模型通常可以充分拟合训练样本，而简单的模型则有助于推广到训练样本以外的实例（泛化能力强）。同时，紧凑的模型参数少、表达能力强但往往求解复杂，而易于求解的模型则往往需要更多参数才能足够表达学习结果。因此在模型选择中一般存在拟合能力与泛化能力之间、求解难度与表达能力之间的折中，但在模型选择方法的设计中通常遵循“越自然越平滑”的平滑性原理[212, 217, 221]、“越简单越有效”的“奥卡姆剃刀（Occam′s razor）”原理[213, 218, 223]、“预测误差越小（泛化能力越强）越好”的实用性原理[214, 216, 219, 221]，力求避免出现对训练数据欠拟合和过拟合这两种极端情况[212, 218, 220-222]。

2）优化准则及其求解（学习算法设计）问题

模型确立后，设计恰当的参数估计方法就是学习算法，包括优化准则及其求解算法两个具体的环节，这是典型的最优化问题。因此，最优化算法是机器学习的重要工具[224-226]。例如，基本的梯度下降法被用于感知机和 CNN 的学习中[99, 102, 105, 154, 227, 228]，随机梯度下降法[229]则是 MLP 学习、BP 算法、强化学习和深度学习的基础性方法[50, 101-105, 230, 231]，而高斯–牛顿法[232-234]则被用于非线性回归、深度学习和 MLP 学习中[234-236]，等等。最优化方法在推动机器学习算法快速发展的同时，自身也因此得到不断进步[237, 238]。

3）算法实现（硬件结构或软件架构）问题

机器学习的算法实现问题，就是实现机器学习算法的硬件电路或软件架构设计问题。例如，感知机、MLP、CNN 通常作为 BP 算法及其改进算法的实现形式[101-105, 228-231]，利用 RBF-NN 既可以实现 BP 算法也可以实现 SVM 学习[125, 204]，正则化网络是正则化学习算法的基本实现形式[212]，而 GP-GPU 则通常作为深度学习算法的神经网络实现形式[150-160]，等等。可见，数据驱动的机器学习算法主要用神经网络实现（参见图 1.1）。

### 2. 机器学习的基本类型和典型算法

按训练数据集中所蕴含信息的情况，机器学习分为有师学习、无师学习和强化学习三种基本类型[208-210, 239]。

首先，若图 1.2 中训练数据集不仅包括实例 $\boldsymbol{x}(n) = [x_1(n), x_2(n), \cdots, x_M(n)]^{\mathrm{T}}$，还同时包含对应的响应 $g(n)$（如函数值、问题答案、模式类别标签等，统称为标签），这时基于训练数据集 $D = \{\boldsymbol{x}(n), \boldsymbol{g}(n)\}$（$n = 1, 2, \cdots, N$）的机器学习称为有师学习或有监督学习（supervised learning）。这里的标签 $g(n)$可以是纯净的（如标准答案），也可以是带噪声的（如参考答案）。对一元函数逼近问题，图 1.4[211]给出了有师学习问题的一般情况。

在图 1.4 中，含噪标签可以表示为

$$\boldsymbol{g} = \mathcal{A} f_0 + \boldsymbol{v}, \tag{1.1}$$

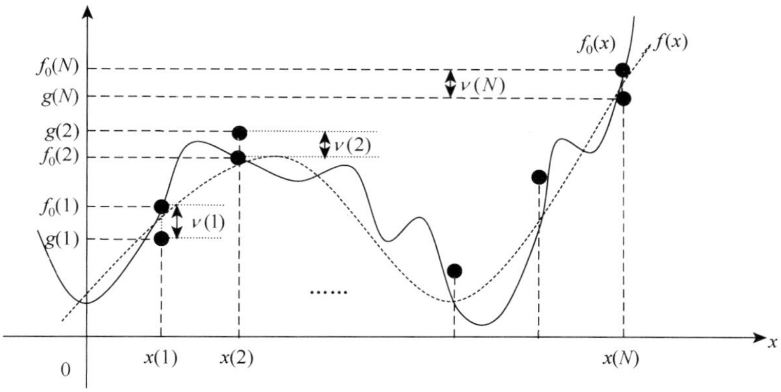


图 1.4　有师学习：函数逼近问题[211]

式中，$\boldsymbol{g}=[g(1), g(2), \cdots, g(N)]^{\mathrm{T}}$，$\boldsymbol{\nu}=[\nu(1), \nu(2), \cdots, \nu(N)]^{\mathrm{T}}$，$\boldsymbol{g}$ 和 $\boldsymbol{\nu}$ 分别为标签向量和标签噪声向量；$\mathcal{A}$ 为数据生成算子（由实例和 $f_0(\boldsymbol{x})$生成标签的过程，如观测时为观测算子，采样时为采样算子等）。

典型的有师学习算法包括回归分析[127, 200, 233, 234]、k-NN 算法[203]、决策树算法[240, 241]、SVM 算法[125, 130, 209]、贝叶斯分类法[242, 243]，等等。有师学习算法主要用于函数逼近和模式分类。

其次，如果图 1.2 中的训练数据集仅仅包括实例而没有对应标签，则基于训练数据集 $D=\{\boldsymbol{x}(n)\}$（$n=1, 2, \cdots, N$）的机器学习称为无师学习或无监督学习（unsupervised learning）。典型的无师学习算法包括源于卡-洛变换（Karhunen-Loève transform，KLT）的主成分分析（principal component analysis，PCA）[244, 245]、线性鉴别分析（linear dicriminant analysis，LDA）[246]、k 均值算法[247]，等等。无师学习算法的优势在于发掘数据的内在结构和相似性，因此往往用于数据降维和聚类分析。

最后，强化学习是介于有师学习与无师学习之间、按学习任务和决策要求来引入并利用奖惩信息进行的学习[3, 121-124, 185, 248, 249]。强化学习是行为主义学派中决策与控制理论和技术发展的产物——学习控制[121, 122, 248, 249]，主要用于训练智能体[133-135, 248, 249]。

需要说明的是，机器学习的这三种基本类型是相对划分的，而以此为基础产生的派生类型（如半监督学习[221, 250, 251]、生成式学习[205, 206]、鉴别学习[252, 253]、主动学习[254-257]、表示学习[160, 218, 258-261]等）则层出不穷，成为机器学习研究和应用中的多个细分方向。

本书重点探讨有师学习中的投影学习算法。

# 1.3　投影学习综述

本节对投影学习理论、算法的研究及应用现状进行综述。

## 1.3.1　联想记忆网络训练中的投影学习

投影学习最早源于数据分析中的最小二乘法，后来在从理论上解决感知机（联想记忆网络）训练问题方面得到学者们的广泛重视。

1. 最小二乘法中的正投影

19 世纪初期高斯和勒让德各自独立发明的最小二乘（least square，LS）法，主要用于估计曲线拟合系数[232, 262]，后来作为一种基本的优化工具，在涉及数据分析、系统辨识、建模预测等问题的众多领域中得到不断应用、完善和推广[263-276]。

尽管如此，最小二乘法的投影形式是于 20 世纪中期在处理离散有限维空间中线性模型参数估计问题时引入的[268]。事实上，对离散有限维空间而言，由因变量 $f_0(\boldsymbol{x})$和 $M$ 个自变量（解释变量，即 $x_1$、$x_2$、…、$x_M$）构成的线性模型为

$$f_0(\boldsymbol{x}) = \sum_{m=1}^{M} a_m x_m = \boldsymbol{x}^{\mathrm{T}}\boldsymbol{a}, \tag{1.2}$$

式中，$\boldsymbol{a} = [a_1, a_2, \cdots, a_M]^{\mathrm{T}}$ 为模型参数向量（未知）；$\boldsymbol{x} = [x_1, x_2, \cdots, x_M]^{\mathrm{T}}$。

对式（1.2）中的 $f_0(\boldsymbol{x})$进行 $N$ 次观测，得到与式（1.1）对应的观测方程：

$$\boldsymbol{g} = \boldsymbol{A}\boldsymbol{a} + \boldsymbol{v}, \tag{1.3}$$

式中，$\boldsymbol{g}$ 和 $\boldsymbol{v}$ 的含义同式（1.1）且假设噪声均值为 0（即无偏移量），而

$$\boldsymbol{A} = [\boldsymbol{x}_1, \boldsymbol{x}_2, \cdots, \boldsymbol{x}_M\ ],\quad \boldsymbol{x}_m = [x_m(1), x_m(2), \cdots, x_m(N)]^{\mathrm{T}},\quad m=1,2,\cdots,M, \tag{1.4}$$

式中，$\boldsymbol{A}$ 为 $N \times M$ 维观测矩阵（数据生成算子 $\mathcal{A}$ 的特殊形式）；$\boldsymbol{x}_m$ 为对第 $m$ 个解释变量的观测向量。

最小二乘法的优化目标函数为

$$J(\boldsymbol{a}) = \frac{1}{2}\sum_{n=1}^{N}[g(n) - f(n)]^2 = \frac{1}{2}\boldsymbol{d}^{\mathrm{T}}\boldsymbol{d}, \tag{1.5}$$

式中，

$$\boldsymbol{d} = \boldsymbol{g} - \boldsymbol{f} = [g(1) - f(1),\quad g(2) - f(2),\quad \cdots,\quad g(N) - f(N)]^{\mathrm{T}}, \tag{1.6}$$

$$f(n) = [\boldsymbol{x}(n)]^{\mathrm{T}}\boldsymbol{a},\quad \boldsymbol{x}(n) = [x_1(n), x_2(n), \cdots, x_M(n)]^{\mathrm{T}},\quad n=1,2,\cdots,N, \tag{1.7}$$

其中，$\boldsymbol{d}$ 为残差（观测值之修正前后之差）向量；$f(n)$为对因变量之第 $n$ 次观测值 $g(n)$进行修正（估计）的结果。

由于式（1.5）中的目标函数为参数向量 $\boldsymbol{a}$ 的二次型函数，其极小值就是最小值，因此若矩阵 $\boldsymbol{A}^{\mathrm{T}}\boldsymbol{A}$ 满秩（其逆存在），则 $\boldsymbol{a}$ 的最小二乘解为

$$\boldsymbol{a}_{\mathrm{LS}} = \underset{\{\boldsymbol{a}\}}{\operatorname{argmin}}\, J(\boldsymbol{a})=(\boldsymbol{A}^{\mathrm{T}}\boldsymbol{A})^{-1}\boldsymbol{A}^{\mathrm{T}}\boldsymbol{g}. \tag{1.8}$$

将式（1.7）中的参数向量 $\boldsymbol{a}$ 用式（1.8）中的估计结果 $\boldsymbol{a}_{\mathrm{LS}}$ 代替，可得到观测向量之修正向量（即模型输出）：

$$\boldsymbol{f} = \boldsymbol{A}(\boldsymbol{A}^{\mathrm{T}}\boldsymbol{A})^{-1}\boldsymbol{A}^{\mathrm{T}}\boldsymbol{g}. \tag{1.9}$$

正是由于 $\boldsymbol{A}(\boldsymbol{A}^{\mathrm{T}}\boldsymbol{A})^{-1}\boldsymbol{A}^{\mathrm{T}}$ 是对称的幂等矩阵（因而是正投影矩阵），所以由式（1.9）可知，模型输出向量 $\boldsymbol{f}$ 是观测向量 $\boldsymbol{g}$ 在超平面 $S$ 上的正投影（垂直投影），如图 1.5 所示。而且，由于这个超平面由 $\boldsymbol{A}$ 的列向量张成（记为 $S = \mathrm{span}\{x_1, x_2, \cdots, x_M\}$），所以最小二乘估计过程就是由解释变量在其所对应的观测空间中对因变量进行“最合理”解释的过程，其合理性体现为残差强度。

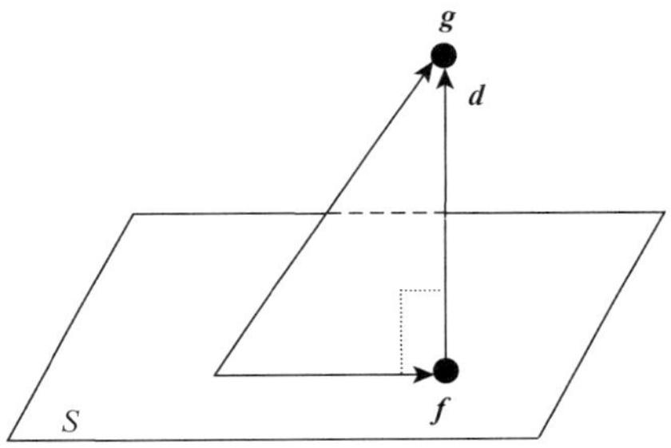


图 1.5　最小二乘法中的正投影原理

另外，最小二乘法既可以推广到非线性模型[265, 269, 270, 274, 276]，也可以推广到解释变量带有观测误差的情形[273]，还可以推广到有多个因变量的情形[275]。

2. 联想记忆网络训练中的投影学习

最小二乘法作为一种基本的优化工具，自然被学者们应用于神经网络参数（连接强度）的估计中，这是 20 世纪 70 年代以后的事[86, 89, 93, 277-283]。事实上，对最早用于线性分类的单层感知机（联想记忆网络）[42, 46]而言，其激励 $\boldsymbol{x}$ 和响应的关系 $f_0(\boldsymbol{x})$可用式（1.2）表示（响应是 $M$ 个激励的线性组合，$f_0(\boldsymbol{x})>0$ 则感知机输出为正类，否则为负类），而式（1.3）则表示观测 $N$ 个时刻（$N$ 个快拍）得到对应含噪响应的过程，由式（1.9）得到响应的估计结果称为近似回想（approximate recall）[283]，式（1.8）给出的权系数则作为联想记忆网络的连接强度参数，而对应网络则记忆了样例激励（训练数据）与对应响应（近似回想）的呼应关系，即式（1.9）。因此，对应的最小二乘准则称为记忆学习准则。

值得指出，当矩阵 $\boldsymbol{A}$ 的列向量（$M$ 个激励的观测向量）线性独立时，矩阵 $\boldsymbol{A}^{\mathrm{T}}\boldsymbol{A}$ 满秩，这时$(\boldsymbol{A}^{\mathrm{T}}\boldsymbol{A})^{-1}\boldsymbol{A}^{\mathrm{T}}$ 就是 $\boldsymbol{A}$ 的穆尔-彭罗斯广义逆（简称广义逆或伪逆）[284-286]，记为 $\boldsymbol{A}^{+}$，即此时 $\boldsymbol{A}^{+}=(\boldsymbol{A}^{\mathrm{T}}\boldsymbol{A})^{-1}\boldsymbol{A}^{\mathrm{T}}$。特别地，当 $\boldsymbol{A}$ 的列向量规范正交时，矩阵 $\boldsymbol{A}^{\mathrm{T}}\boldsymbol{A}$ 为单位阵，式（1.8）和式（1.9）就分别是赫布学习准则给出的连接强度和完整回想（perfect retrieval）[2, 89, 287, 288]。

在一般情况下，$\boldsymbol{A}$ 的列向量可能既不是正交也不是线性独立的，但 $\boldsymbol{A}$ 的广义逆始终存在，这时与式（1.8）和式（1.9）对应，权系数及响应（近似回想）分别取为

$$\boldsymbol{a}_{\mathrm{LS}}=\boldsymbol{A}^{+}\boldsymbol{g} \tag{1.10}$$

和

$$\boldsymbol{f}=\boldsymbol{A}\boldsymbol{A}^{+}\boldsymbol{g}. \tag{1.11}$$

这时权系数向量 $\boldsymbol{a}_{\mathrm{LS}}$ 具有最小范数（最小能量）[286]，而神经网络响应（近似回想）$\boldsymbol{f}$ 依旧是观测结果 $\boldsymbol{g}$ 在 $\boldsymbol{A}$ 的列向量所张成子空间上的正投影，如图 1.5 所示。正因如此，称这一训练算法为投影学习算法，它是联想记忆网络学习的经典方法。

值得指出的是，记忆学习准则是机器学习算法中被广泛使用的准则[86, 89, 93, 104, 218, 283, 287-293]。另外，一旦对应于式（1.3）的观测不含噪声，则 $\boldsymbol{g}=[f_0(1),f_0(2),\cdots,f_0(N)]^{\mathrm{T}}$ 就是由 $N$ 个快拍点对应的完整回想，最小二乘准则就成为最小均方误差（mean squared error，MSE）准则的实现形式（经验形式）[104, 218]。从这个意义上说，记忆学习准则蕴含了 LS 准则和 MSE 准则。此外，对于一般的联想记忆网络，其激励 $\boldsymbol{x}$ 和响应 $f_0(\boldsymbol{x})$（在神经网络中通常称为激励函数）的非线性关系可表示为

$$f_0(\boldsymbol{x})=f_0\left(\sum_{m=1}^{M}a_m x_m+a_0\right)=f_0(\boldsymbol{x}^{\mathrm{T}}\boldsymbol{a}+a_0), \tag{1.12}$$

式中，$a_0$ 为偏移量（如噪声的均值）。一旦 $f_0(\boldsymbol{x})$的形式已知（如已由模型选择确定）而参数未知，则学习目的在于估计偏移量、权系数向量等参数，称为参数模型法，而这时的投影学习算法对应于非线性模型参数估计的最小二乘法[271, 277-282, 290, 291, 293]。

### 1.3.2 最优泛化投影学习

投影法作为记忆学习准则的一种求解算法，也如其他求解算法（如作为非参数模型求解的 BP 算法[101-105, 218]）一样，其中存在的泛化能力局限、正则化改进、最优泛化推广等问题也备受关注。

1. 记忆学习准则的正则化

泛化能力是刻画机器学习性能的重要指标，泛指学习结果的推广能力[28, 99, 294-298]。对记忆学习准则而言，基于训练数据集，无论是使残差平方和最小（LS 准则）还是使误差平方和最小（最小 MSE 准则），都只能确保在观测点（图 1.4 中离散点）给出最优解，难以确保观测点以外（即所有预测点处）的残差或误差最小，因此容易出现过拟合（overfitting）或过学习（overlearning/overtraining）现象[299-310]，限制了学习结果的泛化能力。因此，抑制过拟合现象是提升记忆学习准则和神经网络泛化能力的基本途径，而正则化则作为抑制过拟合的基本手段[212, 221, 310]，成为机器学习算法研究中一个重要课题。

经典的正则化手段是利用先验知识（如学习函数的平滑性）对记忆学习准则进行修正，以此改进泛化能力[212, 311]。事实上，设潜在的未知规则为式（1.12）对应的理想函数 $f_0(\boldsymbol{x})$，而观测方程为式（1.1）。由于学习结果 $f(\boldsymbol{x})$依赖于受误差 $\boldsymbol{v}$ 影响的观测样本（训练数据集），所以 $f(\boldsymbol{x})$往往是随机的。若由训练数据集寻找 $f_0(\boldsymbol{x})$ 而得到 $f(\boldsymbol{x})$的代价函数为 $c(f, f_0)$，则其数学期望被称为风险，记为 $\mathcal{R}=E[c(f, f_0)]$。另外，由于 $f(\boldsymbol{x})$的概率分布属性往往是未知的，所以实际的风险只能由训练数据集（即经验数据）表达，称为经验风险。

对于误差平方代价而言，记忆学习准则对应的经验风险为

$$\mathcal{R}_{\mathrm{MSE\_emp}}(f) \equiv \frac{1}{N}\sum_{n=1}^{N}[f_0(n) - f(n)]^2, \tag{1.13}$$

此即 MSE 准则对应的经验风险，使该风险最小的准则又称为最小均方（least mean square，LMS）准则[104, 218]。

对于残差平方代价，记忆学习准则对应的经验风险为

$$\mathcal{R}_{\mathrm{LS\_emp}}(f) \equiv \frac{1}{N}\sum_{n=1}^{N}[g(n) - f(n)]^2, \tag{1.14}$$

显然与式（1.5）等价（忽略不影响求解结果的常数因子），因而使其最小的准则就是 LS 准则。

若经验风险统一用 $\mathcal{R}_{\mathrm{emp}}(f)$表示，则经典正则化是在 $\mathcal{R}_{\mathrm{emp}}(f)$后附加对求解结果 $f(\boldsymbol{x})$的约束项，得到正则化经验风险为

$$\mathcal{R}_{\mathrm{reg}}(f) = \mathcal{R}_{\mathrm{emp}}(f) + \lambda\varphi(f), \tag{1.15}$$

式中，$\lambda$ 为正则化参数；$\varphi(f)$为正则化约束函数。按平滑性原理[212, 217, 221]，最基本的约束是平滑性约束，即要求求解结果 $f$ 足够平滑。因此典型的正则化（如吉洪诺夫正则化）经验风险为[212, 311]

$$\mathcal{R}_{\mathrm{reg}}(f) = \mathcal{R}_{\mathrm{emp}}(f) + \lambda\|\mathcal{D}f\|^2, \tag{1.16}$$

式中，$\mathcal{D}$ 为微分算子；$\|\cdot\|$为函数 $\mathcal{D}f$ 所属空间（如平方可积函数空间，即能量有限信号空间）的范数。

对于式（1.13）或式（1.14）中的记忆学习准则，式（1.16）中的正则化经验风险就是正则化记忆学习准则，完成其求解可以用多种优化算法，包括一阶梯度下降法（如 BP 算法[101-105]）、二阶梯度下降法（如高斯-牛顿法[232-234, 236]）、格林函数法[312-314]，等等。基于格林函数法的求解结果表明，对于独立观测样本，$f(\boldsymbol{x})$ 隶属于平滑函数空间的 $N$ 维子空间（该子空间由经验数据和微分算子所对应的格林函数共同决定）[212]：

$$f(\boldsymbol{x}) = \sum_{n=1}^{N} \alpha(n) G(\boldsymbol{x}, \boldsymbol{z})\big|_{\boldsymbol{z}=\boldsymbol{x}(n)}, \tag{1.17}$$

式中，格林函数 $G(\boldsymbol{x}, \boldsymbol{z})$是定义在 $R^M \times R^M$ 上的双变量函数（$R$ 表示实数域）。

对于最小二乘准则，吉洪诺夫正则化对应于岭回归（旨在排除解释变量之间的相关性即多重共线性，并以此进行变量选择的回归），这时的正则化参数 $\lambda$ 就是岭回归参数[315-319]。事实上，当解释变量 $x_1$、$x_2$、…、$x_M$ 存在相关关系时，式（1.4）中 $\boldsymbol{A}$ 的列向量（解释变量的观测向量）之间是线性相关的，这时 $\boldsymbol{A}^{\mathrm{T}}\boldsymbol{A}$ 之逆不存在，为此普通最小二乘法不能求解（其数值解不稳定），利用吉洪诺夫正则化方法对式（1.5）进行正则化后，可得[315]

$$J_{\mathrm{reg}}(\boldsymbol{a}) = J(\boldsymbol{a}) + \frac{\lambda}{2}\|\boldsymbol{a}\|^2, \tag{1.18}$$

其求解结果为[315, 319]

$$\boldsymbol{a}_{\mathrm{LS,reg}} = \underset{\{\boldsymbol{a}\}}{\operatorname{argmin}}\, J_{\mathrm{LS,reg}}(\boldsymbol{a}) = (\boldsymbol{A}^{\mathrm{T}}\boldsymbol{A} + \lambda\boldsymbol{I})^{-1}\boldsymbol{A}^{\mathrm{T}}\boldsymbol{g}, \tag{1.19}$$

式中，$\boldsymbol{I}$ 为单位矩阵。于是回归输出（学习结果或神经网络的近似回想）为

$$\boldsymbol{f} = \boldsymbol{A}(\boldsymbol{A}^{\mathrm{T}}\boldsymbol{A} + \lambda\boldsymbol{I})^{-1}\boldsymbol{A}^{\mathrm{T}}\boldsymbol{g}. \tag{1.20}$$

这时，由于$(\boldsymbol{A}^{\mathrm{T}}\boldsymbol{A} + \lambda\boldsymbol{I})^{-1}\boldsymbol{A}^{\mathrm{T}}$ 是矩阵 $\boldsymbol{A}$ 之广义逆 $\boldsymbol{A}^{+}$的一般形式[286]，所以式（1.11）依然成立。因此，岭回归对应的学习算法就是正则化投影学习，且由式（1.16）～式（1.18）可知，岭回归准则就是吉洪诺夫正则化中对求解结果取二阶微分的特例。

除了上述经典正则化，其他正则化（如基于局部随机扰动优化的正则化[320]）手段也能提升记忆学习准则的泛化能力，但由于经验风险准则本身对泛化能力没有保证，这些正则化手段对泛化能力的提升作用非常有限。对于上述经典正则化手段而言，泛化能力的提升程度受到正则化参数 $\lambda$ 影响很大，恰当地估计该参数也是其中一个难题[318, 321]。因此，正则化记忆学习依旧不能确保传统联想记忆网络的最优泛化能力。

2. $\varepsilon$不敏感代价函数与 SVM 学习

在鲁棒估计中，依据噪声向量 $\boldsymbol{v}$ 的概率分布模型来选择代价函数的理论表明[322]：当 $\boldsymbol{v}$ 的概率密度函数呈对称分布时，利用最小最大（min max）策略（在最差噪声分布模型下得到最优泛化估计结果）可以得到绝对残差代价函数，而 $\varepsilon$ 不敏感代价函数是绝对残差代价函数的推广[125]：

$$c(g,f)=|g-f|_{\varepsilon}^{k}=\begin{cases}0, & 若|g-f|^{k}\leqslant\varepsilon\\ |g-f|^{k}-\varepsilon, & 其他\end{cases}$$

或等价地

$$c(g,f)=|g-f|_{\varepsilon}^{k}=\max\left(0,|g-f|^{k}-\varepsilon\right),\tag{1.21}$$

式中，$\varepsilon>0$；$k$ 取 1（对应于 $\varepsilon$ 不敏感绝对残差代价）或 2（对应于 $\varepsilon$ 不敏感残差平方代价）。

对 $\varepsilon$ 不敏感绝对残差代价，对应的经验风险为

$$\mathcal{R}_{\mathrm{emp},\varepsilon}(f)\equiv\frac{1}{N}\sum_{n=1}^{N}|g(n)-f(n)|_{\varepsilon},\tag{1.22}$$

这时，对于式（1.2）中的线性函数 $f_0(\boldsymbol{x})$，由式（1.22）得到与式（1.18）对应的正则化经验风险——正则化 $\varepsilon$ 不敏感绝对残差经验风险为

$$\mathcal{R}_{\mathrm{reg},\varepsilon}(\boldsymbol{a})=\frac{1}{N}\sum_{n=1}^{N}|g(n)-f(n)|_{\varepsilon}+\frac{\lambda}{2}\|\boldsymbol{a}\|^{2},\quad f(n)=[\boldsymbol{x}(n)]^{\mathrm{T}}\boldsymbol{a},\tag{1.23}$$

对该准则求解，得到[125, 209]

$$\boldsymbol{a}=\frac{1}{N}\sum_{n=1}^{N}b_n\boldsymbol{x}(n),\tag{1.24}$$

式中，系数 $b_n(n=1,2,\cdots,N)$只有少量非零（具体数量取决于 $\varepsilon$ 的值），非零系数对应的 $\boldsymbol{x}(n)$称为支持向量，由此得到的学习结果 $f(\boldsymbol{x})$称为支持向量机（SVM）。

对于式（1.12）中的非线性函数 $f_0(\boldsymbol{x})$，通常可以通过非线性变换将 $\boldsymbol{x}$ 映射到高维空间后再应用上述结果，实现非线性模型的 SVM 学习[130]。

可见，基于 $\varepsilon$ 不敏感代价函数的 SVM 学习，其泛化能力要明显优于基于记忆学习准则的传统算法（如 BP 算法）。同时，SVM 的求解最终转换成二次规划问题[125, 209]，其解是唯一的、全局最优的和稀疏的（能以少量样本代表所有样本的特性）。正因如此，SVM 学习一经提出就受到普遍重视，并在模式分类、图像处理、无线信道估计、时间序列预报、目标检测等众多领域得到广泛应用[323-327]。由 SVM 学习派生出来的核方法（kernel method）或核学习（kernel learning），多年来成为模式识别、机器学习、非线性信号处理等领域理论研究和工程应用中的一个热点[328-332]。

在 SVM 学习及其主要改进方法中，由于采用 $\varepsilon$ 不敏感代价函数，所以虽然不存在记忆准则求解典型算法（如 BP 算法）中普遍存在的局部收敛问题，但仍存在以下主要问题。

首先，对大规模样本集学习效率低。由于二次规划所常用的内点算法需要占据过多的内存和计算资源，因而对大规模样本集的学习效率低下。

其次，存在多元最优局限。由于 SVM 学习只对二元问题最优，对于诸如多类别模式分类、多进制相位编码信号检测等多元问题，存在泛化能力优化局限。

另外，增量学习难以完整实现。由于 SVM 学习完全依赖于历史数据（批处理），不支持增量学习——利用新的信息（简称新息）对已有学习结果进行直接修正的在线学习，所以对数据流或流媒体（如音视频）或基于分段/分块处理的信息检索、目标检测、特征选择和分类等不适用。

尽管针对这些问题的改进方法层出不穷[278, 279, 333-342]，但在面对诸多现实应用问题时尚有很大的改进空间，而且 SVM 学习及其改进算法始终未脱离取决于观测空间的经验风险，因此还不是严格意义上的最优泛化学习。

### 3. 数据降维中的最优泛化正投影学习

数据降维即数据压缩，通常指在不丢失或很少丢失有用信息的前提下，缩减数据量或对数据进行重新组织，以降低维数、减少存储空间、提高传输和处理效率的一种技术。

数据压缩的基本依据是数据中普遍存在统计冗余性，通常表现为数据时间相关性或空间相关性。因此，消除或减少相关性是数据压缩算法的基本出发点，而通过数据变换后在变换域编码则是最重要的数据压缩方式之一[343]。通过适当的变换将数据转换为不相关数据，是数据压缩中的一个焦点问题，解决这一问题的经典方法为卡-洛变换（KLT）。鉴于 PCA 作为典型的无师学习和数据降维方法，属于投影学习且与 KLT 有密切的关系[344]，因此这里重点介绍 KLT。

KLT 是最小均方误差意义下的随机正交变换[244, 245]，最早来源于随机过程（随机信号）去相关的霍特林变换[345]，其基本目的在于将随机信号变换成不相关随机序列。事实上，若 $f_0(x)$是定义在（0，$T$）上的一维随机信号（即 $x$ 为标量），寻找（0，$T$）上某标准的确定函数集：

$$\{\psi_n(x)\}_{n=1}^{\infty},\quad \int_0^T \psi_n(x)\overline{\psi_m(x)}\,\mathrm{d}x = \delta_{nm} = \begin{cases} 1, & m = n, \\ 0, & m \neq n, \end{cases} \tag{1.25}$$

式中，$\overline{\bullet}$ 表示复数取共轭。以该函数集对 $f_0(x)$进行离散化得到随机序列 $g(n) = \langle f_0(x), \psi_n(x)\rangle (n = 1, 2, \cdots)$，其中$\langle \bullet, \bullet \rangle$为定义在实数域 $R$ 上之平方可积函数空间 $L^2(R)$中定义的内积。为消除随机序列 $g(n)$的相关性，要求

$$E\left[\{g(n)-E[g(n)]\}\overline{\{g(m)-E[g(m)]\}}\right]=\lambda_n\delta_{nm}\quad n,m=1,2,\cdots \tag{1.26}$$

即序列在不同时刻的样本是不相关的（协方差为零）。同时，要求利用随机序列 $g(n)$和该函数集重构的随机信号 $f(x)$是 $f_0(x)$的最小 MSE 逼近：

$$f(x)=\sum_{n=1}^{N}g(n)\psi_n(x), \tag{1.27}$$

$$\lim_{N\to\infty}\int_0^T[f_0(x)-f(x)]^2\mathrm{d}x=\lim_{N\to\infty}\int_0^T\left[f_0(x)-\sum_{n=1}^{N}g(n)\psi_n(x)\right]^2\mathrm{d}x=0, \tag{1.28}$$

由此得到

$$\int_0^T\int_0^T\psi_n(x_1)\overline{\psi_m(x_2)}C_{f_0}(x_1,x_2)\mathrm{d}x_1\mathrm{d}x_2=\lambda_n\delta_{nm}, \tag{1.29}$$

其中

$$C_{f_0}(x_1,x_2)=E\left[f_0(x_1)\overline{f_0(x_2)}\right]-E[f_0(x_1)]E\left[\overline{f_0(x_2)}\right]. \tag{1.30}$$

式（1.30）是 $f_0(x)$的协方差函数。由式（1.29）可知，需要寻找的正交函数集就是 $f_0(x)$的协方差函数之本征函数集（eigenfunctions）。因此，用随机信号协方差函数的本征函数集对随机信号自身的采样变换

$$g(n)=< f_0(x),\psi_n(x)>,\quad n=1,2,\cdots \tag{1.31}$$

就称为卡-洛变换（KLT）。

由于式（1.28）对所有 $x$（而不是仅限于 $x$ 的观测点）成立，所以对应的最小 MSE 准则是最优泛化准则。同时，由于内积运算式（1.31）对应于正投影，所以由式（1.27）可知：$f(x)$是 $f_0(x)$在空间 $S=\mathrm{span}\{\psi_1(x),\psi_2(x),\cdots,\psi_N(x)\}$上的正投影，这个正投影均方收敛于 $f_0(x)$。因此，式（1.27）称为随机信号的卡-洛展开（Karhunen-Loève expansion，KLE），KLE 是 MSE 意义下的最优泛化正投影学习。

同样地，对于随机序列，用其协方差矩阵的本征向量集（eigenvectors）对其自身进行的采样变换（内积运算）就是离散 KLT，而 PCA 是离散 KLT 的等价形式[344]，因此 PCA 降维算法是 MSE 意义下的最优泛化正投影学习。

尽管 KLE 是 MSE 意义下的最优泛化正投影学习，但由于涉及协方差函数/矩阵的特征分解运算，所以 KLT/PCA 的增量形式难以实现，限制了 KLE 及 KLT/PCA 在有在线学习需要场合的有关应用[346-353]。

#### 4. 基于最优泛化准则的投影学习

以日本学者小川和芬兰学者奥亚为代表的研究人员借助于泛函分析中的投影定理，直接从学习的未知函数所属空间出发，研究了基于最优泛化学习准则的投影学习[354-357]。该准则以泛化误差为最优化目标函数，以向未知函数所属空间的

某一子空间（由模型选择准则确定）的投影为约束，优化准则本身就保证了学习结果的最优泛化能力。同时，相关算法可以解决 SVM 学习及其改进算法中存在的主要问题。如在增量学习方面，借助于正投影容易实现精确增量学习（批处理与自适应处理结果精确一致，以保证学习结果的泛化能力）[358-362]。但在很多实际应用（如阵列信号处理与多天线系统）中，学习结果与学习误差所属空间一般不满足正交关系[363-371]，正投影学习算法对符合这一关系的苛刻要求实际上往往难以满足，相关理论及算法因此尚存足够的提升空间。

本书在详细讨论最优泛化正投影学习理论基础上，对包括斜投影学习在内的改进算法进行探讨，并对算法的典型应用情况进行介绍。

## 1.4 本章小结

作为后续章节的先导和基础，本章在简要介绍人工智能基本概念、发展简史、主要研究领域的基础上，对机器学习基本问题和典型算法进行了概述，并对正则化学习、最优泛化投影学习理论和算法进行了综合论述。

# 第 2 章　投影学习的代数与泛函基础

## 2.1　引　　言

投影学习具有坚实的数学基础，其中代数与泛函中的线性空间理论是其最重要的基础。希尔伯特（Hilbert）空间是泛函分析中最为核心的研究对象，其成熟的理论和方法也是本书中主要算法的基本来源。因此，本章对后续章节要用到的希尔伯特空间的基本概念、基本理论和基本方法进行简要和集中介绍，涉及希尔伯特空间与再生核希尔伯特空间、希尔伯特空间中的线性算子、希尔伯特空间中的框架理论与算子广义逆，等等。在全部介绍中，除对必要结论给出证明的出处外，对一般的理论不进行严格的数学推导。

## 2.2　希尔伯特空间与再生核希尔伯特空间

本节主要对希尔伯特空间的定义、投影定理、再生核希尔伯特空间（reproducing kernel Hilbert space，RKHS）的定义和性质进行概述。

### 2.2.1　希尔伯特空间的定义

泛函分析中涉及多种空间，其中希尔伯特空间是有限维向量空间（即欧几里得空间）向无限维向量（如连续函数）空间的推广，是被赋予线性结构、几何结构且有特殊收敛性质的函数集合。本小节在简单介绍线性空间、度量空间、赋范线性空间、内积空间和巴拿赫空间的基础上给出希尔伯特空间的定义，主要取材于文献[372]～文献[374]。

1. 线性空间

线性空间是被赋予了线性结构的非空集合，具体定义如下。

**定义 2.1（线性空间）** 设 $H$ 是定义在数域 $C$（实数域 $R$ 为其特例）上的非空集合（其元素可以是有限维向量，也可以是连续函数，且在未加说明时默认为连续函数），若同时满足以下条件，则称 $H$ 为 $C$ 上的线性空间，称这些条件为线性结构：

（1）在 $H$ 的元素间定义了一个加法运算“+”，$H$ 对该运算是封闭的，即对任意 $f$，$g\in H$，均存在唯一的 $h\in H$ 使得 $h=f+g$。

（2）在 $C$ 与 $H$ 的元素间定义了一个数乘运算“·”，$H$ 对该运算是封闭的，即对任意 $f\in H$ 和任意 $a\in C$ 均存在唯一的 $g\in H$，有 $g=a\cdot f$，简记为 $g=af$。

（3）对于上述加法运算和数乘运算，均满足如下基本规则（称为线性运算规则）：

①加法交换律，即对任意 $f$，$g\in H$，有 $f+g=g+f$；

②加法结合律，即对任意 $f$，$g$，$h\in H$，有 $(f+g)+h=f+(g+h)$；

③零元素存在，即存在元素 $0\in H$，对于任意 $f\in H$，均有 $0+f=f$；

④负元素存在，即对任意 $f\in H$，均存在 $g\in H$，使得 $f+g=0$；

⑤对任意 $f\in H$ 均有元素 $1\cdot f=f$，其中 $1\in C$（普通数“1”）；

⑥对任意 $a$，$b\in C$ 和任意 $f\in H$，均有 $a(bf)=(ab)f$；

⑦数加分配律，即对任意 $a$，$b\in C$ 和任意 $f\in H$，均有 $(a+b)f=af+bf$；

⑧数乘分配律，即对任意 $a\in C$ 和任意 $f$，$g\in H$，均有 $a(f+g)=af+ag$。

### 2. 度量空间

度量空间是被赋予几何距离的非空集合，具体定义如下。

**定义 2.2（度量空间）** 设 $H$ 是定义在数域 $C$ 上的非空集合，若对任意 $f$，$g\in H$，均有一个实数 $\rho(f,g)$，满足：

（1）非负性，即 $\rho(f,g)\geqslant 0$，且 $\rho(f,g)=0$ 当且仅当 $f=g$；

（2）三角不等式成立，即对任意 $h\in H$，均有 $\rho(f,g)\leqslant\rho(f,h)+\rho(g,h)$。

则所定义的 $\rho(f,g)$ 称为 $H$ 中的一个距离测度，定义了距离测度的集合称为度量空间。

需要说明的是，线性空间不一定是度量空间，度量空间也不一定是线性空间。

### 3. 赋范线性空间

赋范线性空间是其元素被赋予了大小测度（范数）的线性空间，具体定义如下。

**定义 2.3（赋范线性空间）** 设 $H$ 是定义在数域 $C$ 上的线性空间，若对任意 $g\in H$ 均有一个实数 $p(g)$，满足：

（1）非负性，即 $p(g)\geqslant 0$，且 $p(g)=0$ 当且仅当 $g=0$；

（2）伸缩性，即任意 $a\in C$，均有 $p(ag)=|a|g$；

（3）三角不等式成立，即对任意 $f\in H$，均有 $p(f+g)\leqslant p(f)+p(g)$。

则所定义的 $p(g)$ 称为 $H$ 中的一个范数，简记为 $\|g\|$，而定义了范数的线性空间称赋范线性空间。

显然，赋范线性空间一定是度量空间（其距离可由范数诱导而得）。

4. 内积空间

内积空间是其元素间被赋予几何角度关系（内积）的线性空间，具体定义如下。

**定义 2.4（内积空间）** 设 $H$ 是定义在数域 $C$ 上的线性空间，若对任意 $f$，$g\in H$ 均有一个对应的数 $<f, g>\in C$，满足：

（1）共轭对称性，即 $<f,g>=\overline{<g,f>}$；

（2）对第一变元存在线性性质，即任意 $a$，$b\in C$ 和任意 $h\in H$ 均有

$$<af+bg, h> = a<f, h> + b<g, h>;$$

（3）非负性，即对任意 $g\in H$，均有 $<g, g>\geqslant 0$。

则所定义的 $<\cdot,\cdot>$ 称为 $H$ 中的内积，而定义了内积的线性空间称内积空间。

显然，内积空间一定是赋范线性空间（其范数可由内积诱导而得），而 $H$ 中若两元素内积为零则称二者正交。

5. 巴拿赫空间与希尔伯特空间

巴拿赫空间和希尔伯特空间均与柯西列和完备空间的概念有密切联系，有关定义如下。

**定义 2.5（柯西列）** 设 $H$ 是定义在数域 $C$ 上的度量空间，$\{g_n\}(n=1,2,\cdots)$ 是 $H$ 中的点列。若对任意 $\varepsilon>0$，存在自然数 $N(\varepsilon)>0$，当自然数 $n$，$m\geqslant N(\varepsilon)$时，均有 $\rho(g_n,g_m)<\varepsilon$，则称 $\{g_n\}$ 为 $H$ 中的柯西（Cauchy）列或基本列。

**定义 2.6（完备空间）** 设 $H$ 是定义在数域 $C$ 上的度量空间，若其中每个柯西列都收敛，则该度量空间称为完备空间。

**定义 2.7（巴拿赫空间）** 完备的赋范线性空间称为巴拿赫（Banach）空间。

**定义 2.8（希尔伯特空间）** 由内积导出范数且完备的内积空间称为希尔伯特空间。

可见，希尔伯特空间是被赋予了线性结构和几何结构（距离、范数、内积）的函数集合。同时，对于无限维空间而言，由于收敛性具有重要的价值，所以完备性被引入到希尔伯特空间。典型的希尔伯特空间有平方可积数列空间 $l^2$、平方可积函数空间 $L^2$[372]。

泛函分析中上述各种空间的相互关系如图 2.1[375]所示。

### 2.2.2　投影定理

希尔伯特空间中的投影定理是本书核心算法的基石。由于该定理涉及正交补的概念，因此在给出定理前先定义正交补。

**定义 2.9（正交补）** 设 $H$ 是定义在数域 $C$ 上的内积空间，$S$ 是 $H$ 的一个子集，则 $S$ 的正交补定义为：$S^{\perp} = \{f\in H: <f, g> = 0, \forall g\in S\}$。

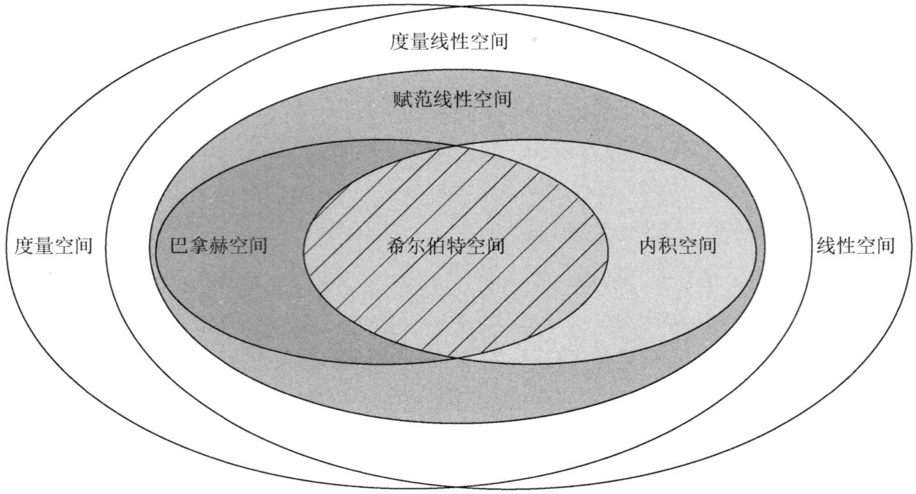


图 2.1　泛函分析中多种空间的相互关系[375]

**定理 2.1（投影定理）** 设 $H$ 是定义在数域 $C$ 上的希尔伯特空间，$S$ 是 $H$ 的一个闭子空间，则对任意 $g\in H$，存在唯一的 $f\in S$ 和 $d\in S^{\perp}$，使得 $g=f+d$，且 $\|g\|^2=\|f\|^2+\|d\|^2$。

该定理的证明可参见文献[373]，而示意图如图 1.5 所示。

希尔伯特空间中的投影定理是欧几里得空间中的投影定理（对二维空间即平面而言，就是勾股定理）的推广，为便于区分，希尔伯特空间中的投影定理有时被称为商高定理。

### 2.2.3　再生核希尔伯特空间

再生核希尔伯特空间是其所有元素（函数）及每个元素的所有取值均可通过对应的再生核得到的希尔伯特空间。本小节在简单介绍再生核的定义及其基本性质的基础上，给出再生核希尔伯特空间的基本性质，主要取材于文献[372]～文献[374]、文献[376]、文献[377]。

1. 再生核及其基本性质

再生核是希尔伯特空间中可能存在的双变量函数，具有再生性，有关定义和性质如下。

**定义 2.10（再生核及其再生性）** 设 $H$ 是定义在抽象集 $\Omega$（可以是实数域、复数域、有限维向量集等）上的希尔伯特空间，对任意 $s$，$t\in\Omega$，若存在定义于 $\Omega\times\Omega$ 上的双变量函数 $k(s, t)$，且同时满足以下条件，则称 $k(s, t)$为 $H$ 的再生核：

（1）元素再生性，即固定 $s$ 则存在 $f(t)=k(s,t)\in H$；

（2）元素值再生性，即对任意 $g\in H$，有 $g(t)=\langle g(s), k(s,t)\rangle_s$（对 $s$ 求内积即得 $g$ 在 $t$ 点的值）。

存在再生核的希尔伯特空间就是再生核希尔伯特空间（RKHS），其再生核有如下基本性质：

（1）唯一性，如果 $H$ 具有再生核 $k(s,t)$，则此再生核是唯一的；

（2）共轭对称性，对于任意 $s$，$t\in\Omega$，均有 $k(s,t)=\overline{k(t,s)}$；

（3）非负性，对于任意 $s\in\Omega$ 均有 $k(s,s)\geqslant 0$；

（4）非负定性，即对于任意 $s_1, s_2, \cdots, s_n\in\Omega$ 以及任意复数 $a_1, a_2, \cdots, a_n$，均有 $\sum_{i=1}^{n}\sum_{j=1}^{n}\overline{a_i}a_j k(s_i,s_j)\geqslant 0$；

（5）柯西不等式成立，即对任意 $s$，$t\in\Omega$，均有 $|k(s,t)|^2\leqslant k(s,s)k(t,t)$。

2. 再生核希尔伯特空间的基本性质

定义在抽象集 $\Omega$ 上的再生核希尔伯特空间 $H$ 有如下性质：

（1）$H$ 具有再生核的充要条件是，对于任意 $t\in\Omega$ 和任意 $f\in H$，映射关系 $f(t)=\langle f(s), g_t(s)\rangle_s$ 均有界，称其中的 $g_t(s)=k(s,t)$ 为 $f$ 在 $t$ 处取值的表示子；

（2）设 $H_1$ 和 $H_2$ 均为在 $\Omega$ 上定义的 RKHS，对应的内积分别为 $\langle\cdot,\cdot\rangle_1$ 和 $\langle\cdot,\cdot\rangle_2$，对应的再生核分别为 $k_1(s,t)$ 和 $k_2(s,t)$。若对任意 $s, t\in\Omega$ 均有 $k_1(s,t)=k_2(s,t)$，那么 $H_1=H_2$，且对于任意 $f, g\in H_1=H_2$，均有 $\langle f,g\rangle_1=\langle f,g\rangle_2$；

（3）由于再生核是唯一的，所以再生核希尔伯特空间与再生核是一一对应的；

（4）若 $k(s,t)$ 是 $\Omega$ 上的非负定函数，则以 $k(s,t)$ 为再生核可在 $\Omega$ 上构造一个希尔伯特空间 $H$；

（5）若 $H_1$ 和 $H_2$ 均为定义在 $\Omega$ 上的 RKHS，对应的范数分别为 $\|\cdot\|_1$ 和 $\|\cdot\|_2$，对应的再生核分别为 $k_1(s,t)$ 和 $k_2(s,t)$，则 $k_1+k_2$ 是所有形如 $g=g_1+g_2$（其中 $g_1\in H_1, g_2\in H_2$）的函数所构成希尔伯特空间 $H$ 的再生核（这样的 $H$ 称为 $H_1$ 和 $H_2$ 的直和，记为 $H=H_1\dotplus H_2$），且 $H$ 的范数为 $\|g\|^2=\min\left[\|g_1\|_1^2+\|g_2\|_2^2\right]$（这一性质可推广到多个空间直和的情形）；

（6）若 $H_1$ 和 $H_2$ 均为定义在 $\Omega$ 上的 RKHS，对应再生核分别为 $k_1(s,t)$ 和 $k_2(s,t)$，则直积（笛卡儿积）$H=H_1\otimes H_2$ 的再生核为 $k(s_1,s_2,t_1,t_2)=k_1(s_1,t_1)k_2(s_2,t_2)$。

## 2.3　希尔伯特空间中的线性算子

在线性代数中，欧几里得空间之间可以通过线性变换建立联系，而矩阵是线性变换的基本形式。欧几里得空间推广到希尔伯特空间后，矩阵理论自然被推广到线性算子理论，以作为研究希尔伯特空间之间关系的基本工具。本节在介绍希尔伯特空间中线性泛函、有界线性算子等有关概念和基本理论的基础上，重点讨论投影算子。

## 2.3.1　希尔伯特空间中的线性泛函与线性算子

希尔伯特空间与数域之间的映射关系用线性泛函刻画，而希尔伯特空间之间的映射关系用算子表征。本小节简要介绍希尔伯特空间中线性泛函与线性算子的有关概念和基本表示形式，主要取材于文献[372]、文献[373]、文献[378]。

1. 线性泛函与线性算子

线性泛函与线性算子是希尔伯特空间中元素（“点”）的函数，有关定义如下。

**定义 2.11（线性泛函）** 设 $H$ 是定义在抽象集 $\Omega$ 上的希尔伯特空间，$S$ 是 $H$ 的子集，则从 $S$ 到复数集 $C$ 的映射关系 $\mathcal{A}$：$\forall g \in S \to \mathcal{A}(g) = \mathcal{A}g \in C$，称为定义在 $H$ 上之定义域(domain)为 $S$ 的泛函；若 $\mathcal{A}$ 还满足线性性质：$\forall g_1, g_2 \in S$ 和 $\forall a_1, a_2 \in C$，均有 $\mathcal{A}(a_1g_1 + a_2g_2) = a_1\mathcal{A}(g_1) + a_2\mathcal{A}(g_2)$成立，则称 $\mathcal{A}$ 为定义在 $H$ 上之定义域为 $S$ 的线性泛函。

**定义 2.12（线性算子及其值域和零空间）** 设 $H_1$ 和 $H_2$ 均为定义在抽象集 $\Omega$ 的希尔伯特空间，对应的内积分别为 $<\cdot, \cdot>_1$ 和 $<\cdot, \cdot>_2$，$S_1$ 为 $H_1$ 子集，则映射关系 $\mathcal{A}$：$\forall g \in S_1 \to \mathcal{A}(g) = \mathcal{A}g \in H_2$，被称为定义域为 $S_1$ 的从 $H_1$ 到 $H_2$ 的算子，满足线性性质的算子 $\mathcal{A}$ 称为线性算子，而 $H_1$ 和 $H_2$ 分别称为 $\mathcal{A}$ 的原像空间(preimage space）和像空间（image space）。

对于线性算子 $\mathcal{A}$，像空间中满足以下关系的集合 $S_2$ 称为其值域（range）：$S_2 = \{g \in H_2: g = \mathcal{A}f, \forall f \in S_1\}$；原像空间中满足以下关系的集合 $Z$ 称为其零空间(null-space 或 kernel)：$Z = \{f \in H_1: \mathcal{A}f = 0\}$。为书写方便，线性算子 $\mathcal{A}$ 的定义域、值域和零空间分别记为 $D(\mathcal{A})$、$R(\mathcal{A})$和 $N(\mathcal{A})$。

显然，在定义 2.12 中，若 $H_2 = C$（复数集），则线性算子就是线性泛函。所以，线性泛函是线性算子的特例。

另外，有如下几种常被引用的特殊算子：

（1）对任意 $f \in S$，均有 $\mathcal{A}f = f$ 的算子 $\mathcal{A}$ 称为单位算子，特记为 $I$；

（2）对任意 $g \in H_2$，至多存在一个 $f \in H_1$ 满足关系 $g = \mathcal{A}f$ 的算子 $\mathcal{A}$ 称为一对一算子或单射（injective）算子；

（3）对任意 $g \in H_2$，至少存在一个 $f \in H_1$ 满足关系 $g = \mathcal{A}f$ 的算子 $\mathcal{A}$ 称为满射(surjective）算子；

（4）对任意 $g\in H_2$，存在唯一一个 $f\in H_1$ 满足关系 $g=\mathcal{A}f$ 的算子 $\mathcal{A}$ 称为双射（surjective）算子（既单射又满射的算子）。

**定义 2.13（有界线性算子及其范数）** 对从希尔伯特空间 $H_1$ 到 $H_2$ 上的线性算子 $\mathcal{A}$，若对于任意 $f\in H_1$ 均存在一正实数 $a$，使得 $\|\mathcal{A}f\|_2\leqslant a\|f\|_1$ 成立，则称 $\mathcal{A}$ 为有界线性算子。对线性算子而言，有界与连续是等价的（文献[373]，引理 4.1）。

有界线性算子 $\mathcal{A}$ 的范数定义为：$\|\mathcal{A}\|=\sup\{\|\mathcal{A}f\|_2: \|f\|_1\leqslant 1\}$，其中“sup”表示取上确界（即取最小的上界）。

说明：由于求内积和取范数都存在对应的操作对象，而内积和范数均对应于操作对象所属空间。为此，为简化起见，除特别需要标注的情况外，后续求内积和取范数时不再用下标标明所属空间。

**定义 2.14（线性算子的伴随算子）** 对从希尔伯特空间 $H_1$ 到 $H_2$ 上的线性算子 $\mathcal{A}$，一定存在一个唯一的从 $H_2$ 到 $H_1$ 上的对应算子 $\mathcal{A}^*$，满足：对任意 $f\in H_1$ 和任意 $g\in H_2$, $\langle\mathcal{A}f, g\rangle=\langle f, \mathcal{A}^*g\rangle$ 成立。称 $\mathcal{A}^*$ 为 $\mathcal{A}$ 的伴随算子（adjoint operator）；若 $\mathcal{A}^*=\mathcal{A}$，则称 $\mathcal{A}$ 为自伴算子。

该定义及相应证明参见文献[373]中定理 6.1。

## 2. 里斯表示定理与诺伊曼-沙滕积

里斯（Riesz）表示定理给出线性泛函的基本表示形式，而用诺伊曼-沙滕（Neumann-Schatten）积则可以表示线性算子，简述如下。

1）里斯表示定理及其几何意义

**定理 2.2（里斯表示定理）** 设 $\mathcal{A}$ 是从希尔伯特空间 $H$ 到数域 $C$ 上的任意泛函，则对任意 $f\in H$，都存在唯一的 $g\in H$，使得 $\mathcal{A}f=\langle f, g\rangle$，且 $\|\mathcal{A}\|=\|g\|$。

该定理的证明参见文献[372]。

由于内积表示正投影（$\langle f, g\rangle$ 是 $f$ 在 $g$ 上的正投影），所以里斯表示定理的几何意义为：希尔伯特空间上的任意泛函都可以用一个对应的投影来表示。

2）诺伊曼-沙滕积与线性算子的表示

**定义 2.15（诺伊曼-沙滕积）** 对定义于同一抽象集 $\Omega$ 上的希尔伯特空间 $H_1$ 和 $H_2$，以及任意 $\psi$，$f\in H_1$ 和任意 $\varphi\in H_2$，定义诺伊曼-沙滕积 $(\cdot\otimes\bar{\cdot})$ 为[378]：

$$(\varphi\otimes\bar{\psi})f=\langle f, \psi\rangle\varphi. \tag{2.1}$$

由这一定义可知，诺伊曼-沙滕积是一从 $H_1$ 到 $H_2$ 上的线性算子，且有如下性质[378]：

（1）$(\varphi\otimes\bar{\psi})$ 的界及值域维数：$\|(\varphi\otimes\bar{\psi})\|=\|\varphi\|\|\psi\|$，$(\varphi\otimes\bar{\psi})$ 的值域维数为 1 或 0;

（2）伴随性：$(\varphi\otimes\bar{\psi})^*=(\psi\otimes\bar{\varphi})$；

（3）对任意复数 $\lambda$，有：$(\lambda\varphi\otimes\bar{\psi})=\lambda(\varphi\otimes\bar{\psi})$, $(\varphi\otimes\overline{\lambda\psi})=\bar{\lambda}(\varphi\otimes\bar{\psi})$；

（4）对任意$\varphi_1$，$\varphi_2 \in H_2$，有：$([\varphi_1+\varphi_2]\otimes\overline{\psi})=(\varphi_1\otimes\overline{\psi})+(\varphi_2\otimes\overline{\psi})$；

（5）对任意$\psi_1$，$\psi_2 \in H_1$，有：$(\varphi\otimes\overline{[\psi_1+\psi_2]})=(\varphi\otimes\overline{\psi_1})+(\varphi\otimes\overline{\psi_2})$；

（6）对任意$\psi_1$，$\psi_2 \in H_1$，$\varphi_1$，$\varphi_2 \in H_2$，有：$(\psi_1\otimes\overline{\varphi_1})(\varphi_2\otimes\overline{\psi_2})=<\varphi_2,\varphi_1>(\psi_1\otimes\overline{\psi_2})$；

（7）对$H_2$上的任意线性算子$\mathcal{G}$，有：$\mathcal{G}(\varphi\otimes\overline{\psi})=(\mathcal{G}\varphi\otimes\overline{\psi})$；

（8）对$H_1$上的任意线性算子$\mathcal{A}$，有：$(\varphi\otimes\overline{\psi})\mathcal{A}=(\varphi\otimes\overline{\mathcal{A}^*\psi})$。

利用诺伊曼-沙滕积可以表示希尔伯特空间中的线性算子，这在后续讨论中经常用到。

### 2.3.2　希尔伯特空间中的投影算子

希尔伯特空间中的投影算子是映射到其自身的一种特殊线性算子。本小节在简要介绍希尔伯特空间分解的基础上，重点介绍投影算子，主要取材于文献[372]～文献[374]。

1. 希尔伯特空间的分解

空间的分解实际上就是在保留原空间结构的同时，将该空间对应的集合进行某种划分（分为多个除零元素外彼此没有公共元素的子集）而得到多个子空间的过程。有关定义如下：

**定义 2.16（希尔伯特空间的分解与补空间）** 设$H$是定义在抽象集$\Omega$上的希尔伯特空间，$S$是$H$的一个子空间（依旧是希尔伯特空间），$S$的分解是指将集合$S$划分成多个满足如下条件的子集$S_i(i=1,2,\cdots,N)$的过程：

（1）对任意$i\neq j$，$S_iS_j=\{0\}$，这里的0为$H$中的零元素（亦即$S$中的零元素）；

（2）对任意$f\in S$，存在唯一的$g_i\in S_i$，使得$f=g_1+g_2+\cdots+g_N$成立；

（3）对任意$i=1,2,\cdots,N$，$S_i$均保留了$S$的所有结构（代数结构、几何结构和完备性）。

这时称$S_i(i=1,2,\cdots,N)$为$S$的直和分解，记为$S=S_1\dotplus S_2\dotplus\cdots\dotplus S_N$。进一步，若对任意$i\neq j$，$S_i$与$S_j$都正交（$S_i$中任意元素与$S_j$中任意元素均正交），则称$S_i(i=1,2,\cdots,N)$为$S$的正交分解，记为$S=S_1\oplus S_2\oplus\cdots\oplus S_N$。

特别地，当$N=2$时，称$S_1$为$S_2$在$S$中的补空间，反之亦然；若还满足$S_1$与$S_2$正交，则$S_1$为$S_2$在$S$中的正交补，反之亦然。

显然，投影定理（定理2.1）也可用希尔伯特空间$H$的正交分解表述[374]。另外，同一个希尔伯特空间，可以存在多种不同的分解。

2. 希尔伯特空间中的投影算子

希尔伯特空间中投影算子的定义及有关性质如下：

**定义 2.17（投影算子）** 设 $H$ 是定义在抽象集 $\Omega$ 上的希尔伯特空间，$S$ 是 $H$ 的一个子空间，$U$ 是 $S$ 在 $H$ 中的补空间（即有 $H=S\dotplus U$），则对任意 $g\in H$，存在唯一的 $f\in S$ 和唯一的 $d\in U$，使得 $g=f+d$ 成立，则称 $f$ 是 $g$ 的沿 $U$ 向 $S$ 的投影，而将每个 $g\in H$ 都映射到其在 $S$ 上之投影 $f\in S$ 的算子 $\mathcal{P}_{S,U}$ 称为 $H$ 中沿 $U$ 向 $S$ 的投影算子（简称投影算子）。

希尔伯特空间中的投影算子 $\mathcal{P}_{S,U}$ 有如下性质：

（1）$\mathcal{P}_{S,U}$ 是有界线性算子；

（2）$\mathcal{P}_{S,U}$ 是幂等算子，即有 $\mathcal{P}_{S,U}\mathcal{P}_{S,U}=(\mathcal{P}_{S,U})^2=\mathcal{P}_{S,U}$。

可以证明，任何满足幂等关系的有界线性算子都是投影算子，而满足自伴关系的投影算子都是正投影算子（对应于 $U=S^{\perp}$ 的情况），反之亦然[372-374]。

作为特例，以下在定义标准正交系的基础上，利用诺伊曼-沙滕积表示正投影算子。

**定义 2.18（标准正交系与算子的迹）** 设 $H$ 是定义在抽象集 $\Omega$ 上的希尔伯特空间，点列 $\{\psi_1,\psi_2,\cdots\}$ 作为 $H$ 的子集，如果满足 $\|\psi_i\|=1(i=1,2,\cdots)$ 且 $<\psi_i,\psi_j>=0(j\neq i)$，则称该子集是规范正交系或标准正交系。

基于 $H$ 中的标准正交系 $\{\psi_1,\psi_2,\cdots\}$，$H$ 中有界线性算子 $\mathcal{A}$ 的迹定义为

$$\mathrm{tr}(\mathcal{A})=\sum_i<\mathcal{A}\psi_i,\psi_i>, \tag{2.2}$$

特别地，若 $S$ 是 $H$ 的子空间，且由标准正交系 $\{\psi_1,\psi_2,\cdots,\psi_N\}$ 张成（即 $S=\mathrm{span}\{\psi_1,\psi_2,\cdots,\psi_N\}$），则从 $H$ 向 $S$ 上的正投影算子 $\mathcal{P}_S$ 可以表示为

$$\mathcal{P}_S=\sum_{i=1}^{N}\left(\psi_i\otimes\overline{\psi_i}\right). \tag{2.3}$$

借助于正投影算子，由式（1.31）对应 KLT 给出重构信号的关系式（1.27）（即 KLE）可表示为 $f(x)=\mathcal{P}_S f_0(x)$，即 $f(x)$ 是 $f_0(x)$ 在 $S$ 上的正投影。

## 2.4　希尔伯特空间中的框架与算子广义逆

作为代数与泛函中两个重要的基本理论，希尔伯特空间中的框架理论[374, 379-381]和算子广义逆理论[382-385]及其相互关系[386]在后续章节中有重要作用，本节对此进行简单介绍。

### 2.4.1　希尔伯特空间中的框架

框架（frame）理论为在原像空间中寻找原像提供了一种基本工具，该理论建

立在稠密空间、绍德尔（Schauder）基、贝塞尔（Bessel）序列等概念的基础上。本小节对这些内容进行简要介绍。

1. 稠密空间与绍德尔基

**定义 2.19（稠密子集、绍德尔基和贝塞尔序列）** 设 $H$ 是定义在抽象集 $\Omega$ 上的希尔伯特空间，其中的函数族（点列）$F=\{\psi_1, \psi_2, \cdots\}$ 作为 $H$ 的子集，若对任意 $f\in H$ 都存在一常数集 $\{a_1, a_2, \cdots, a_N\}$，使得

$$g_N=\sum_{n=1}^{N} a_n \psi_n \xrightarrow{N\to\infty} f \tag{2.4}$$

成立，则称 $F=\{\psi_1, \psi_2, \cdots\}$ 是 $H$ 的稠密子集。

进一步，若对任意 $f\in H$ 都存在唯一数集 $\{a_1, a_2, \cdots\}$，使得

$$f=\sum_{n=1}^{\infty} a_n \psi_n \tag{2.5}$$

成立，则称 $F=\{\psi_1, \psi_2, \cdots\}$ 是 $H$ 的绍德尔基。

另外，若对任意 $f\in H$ 都有

$$\sum_{n=1}^{\infty}\left|< f, \psi_n >\right|^2 < \infty, \tag{2.6}$$

则称 $\{\psi_1, \psi_2, \cdots\}$ 为贝塞尔序列。

2. 框架与框架算子

**定义 2.20（框架）** 设 $H$ 是定义在抽象集 $\Omega$ 上的希尔伯特空间，若其中点列 $F=\{\psi_1, \psi_2, \cdots\}$ 满足条件：对任意 $f\in H$，存在两个常数 $A$ 和 $B(0<A\leqslant B<\infty)$，使得

$$A\|f\|^2 \leqslant \sum_{n=1}^{\infty}\left|< f, \psi_n >\right|^2 \leqslant B\|f\|^2 \tag{2.7}$$

成立，则称 $F$ 是 $H$ 的框架。其中，$A$ 和 $B$ 称为框架边界。若 $A=B$，则称 $F$ 为紧致框架；若 $F$ 还是 $H$ 的绍德尔基，则称 $F$ 为 $H$ 的里斯基。

一般而言，框架不是正交基，而是函数冗余表示的一种工具。例如对二维欧几里得空间（平面）而言，$H=R^2$，任意向量可用三个向量表示：$\boldsymbol{e}_1=(0, 1)$、$\boldsymbol{e}_2=\left(-\sqrt{3}/2, -1/2\right)$ 和 $\boldsymbol{e}_3=\left(\sqrt{3}/2, -1/2\right)$，其模均为 1，相互间的夹角为 120°，即向量族 $\{\boldsymbol{e}_k\}_{k=1}^3$ 不相关也不正交，如图 2.2[380]所示。

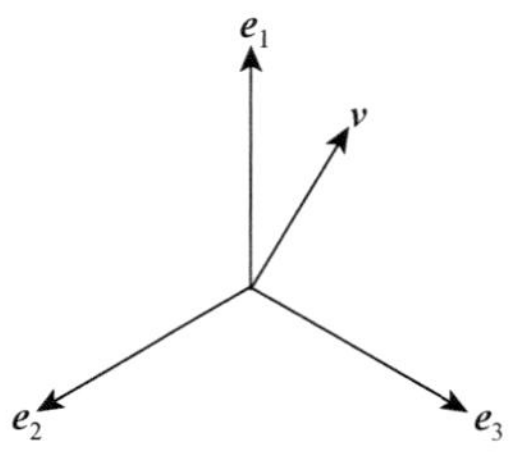

图 2.2　平面中用三个向量 $\boldsymbol{e}_1$、$\boldsymbol{e}_2$ 和 $\boldsymbol{e}_3$ 表示一个任意向量 $\boldsymbol{v}$[380]

在图 2.2 中，$\boldsymbol{v}=(v_1,v_2)$ 为 $R^2$ 中的任意向量，显然有

$$\sum_{k=1}^{3}\left|<\boldsymbol{v},\boldsymbol{e}_k>\right|^2=\left|v_2\right|^2+\left|-\frac{\sqrt{3}}{2}v_1-\frac{1}{2}v_2\right|^2+\left|\frac{\sqrt{3}}{2}v_1-\frac{1}{2}v_2\right|^2=\frac{3}{2}\|\boldsymbol{v}\|^2, \tag{2.8}$$

因此有 $B/A=3/2$，$\{\boldsymbol{e}_k\}_{k=1}^{3}$ 在表示 $R^2$ 中的任意向量时有一个向量是多余的。

事实上，里斯基就是线性独立的点列，在表示函数时无冗余；进一步，对规范化（范数为 1）的里斯基，在框架边界若 $A=B=1$ 成为标准正交基[380]。

**定义 2.21（框架算子）** 设 $F=\{\psi_1,\psi_2,\cdots\}$是希尔伯特空间 $H$ 的一个框架，则框架算子可定义如下：

（1）从 $H$ 到 $H$（实际上是从 $H$ 到点列 $F$ 之所有元素所张成的子空间，即 $S=\mathrm{span}\{\psi_1,\psi_2,\cdots\}$）上的线性映射，即

$$S=\sum_{n=1}^{\infty}\left(\psi_n\otimes\overline{\psi_n}\right),\quad Sf\equiv\sum_{n=1}^{\infty}<f,\psi_n>\psi_n,\forall f\in H, \tag{2.9}$$

这样定义的框架算子 $S$ 是有界、可逆的[380, 381]。

（2）从 $H$ 到 $l^2$ 空间的线性映射，即

$$\mathcal{A}: H\to l^2,\quad c_n=(\mathcal{A}f)_n\equiv<f,\psi_n>,\forall f\in H,\quad \sum_{n=1}^{\infty}\left|c_n\right|^2<\infty, \tag{2.10}$$

这样定义的框架算子 $\mathcal{A}$，其伴随算子 $\mathcal{A}^*$为

$$\mathcal{A}^*:\ l^2\to H,\quad \mathcal{A}^*c\equiv\sum_{n=1}^{\infty}c_n\psi_n,\quad \forall c=[c_1,c_2,\cdots]\in l^2. \tag{2.11}$$

显然，由于

$$\mathcal{A}^*\mathcal{A}:\ H\to H,\quad (\mathcal{A}^*\mathcal{A})f\equiv\sum_{n=1}^{\infty}<f,\psi_n>\psi_n,\quad \forall f\in H, \tag{2.12}$$

所以两种定义给出的框架算子有如下关系：

$$S=\mathcal{A}^*\mathcal{A}. \tag{2.13}$$

3. 对偶框架、重构定理与 RKHS 中再生核的构造[380-382]

**定义 2.22（对偶框架）** 设 $F=\{\psi_1,\psi_2,\cdots\}$是希尔伯特空间 $H$ 中边界分别为 $A$ 和 $B$ 的框架，则由式（2.9）定义的框架算子 $S$ 之逆算子所得子列

$$\widetilde{F}=\{\widetilde{\psi_n}\}_{n=1}^{\infty}:\ \widetilde{\psi_n}=S^{-1}\psi_k=(\mathcal{A}^*\mathcal{A})^{-1}\psi_n \tag{2.14}$$

称为 $F$ 的对偶框架（或共轭框架），$\widetilde{F}$ 对应的框架边界分别为 $B^{-1}$ 和 $A^{-1}$。

若 $F$ 为 $H$ 的里斯基，则框架与对偶框架满足双正交关系，即有

$$<\psi_{\mathrm{m}},\widetilde{\psi_n}>=\delta_{mn}. \tag{2.15}$$

其中，$\delta_{mn}$ 如式（1.25）所示。也就是说，这时 $F$ 和 $\widetilde{F}$ 为 $H$ 的双正交系。

**定理 2.3（重构定理）** 对希尔伯特空间 $H$ 中边界分别为 $A$ 和 $B$ 的框架

$F=\{\psi_n\}_{n=1}^{\infty}$，其对偶框架为 $\widetilde{F}=\{\widetilde{\psi_n}\}_{n=1}^{\infty}$，则任意 $f\in S=\mathrm{span}\{\psi_1,\psi_2,\cdots\}\subset H$ 都可精确重构为

$$f=\sum_{n=1}^{\infty}<f,\widetilde{\psi_n}>\psi_n=\sum_{n=1}^{\infty}<f,\psi_n>\widetilde{\psi_n}\Leftrightarrow f=\mathcal{P}_S f,\mathcal{P}_S=\sum_{n=1}^{\infty}\left(\psi_n\otimes\overline{\widetilde{\psi_n}}\right). \tag{2.16}$$

特别地，对紧致框架有

$$f=\frac{1}{A}\sum_{n=1}^{\infty}<f,\psi_n>\psi_n\Leftrightarrow f=\mathcal{P}_S f,\quad \mathcal{P}_S=\sum_{n=1}^{\infty}\left(\psi_n\otimes\overline{\psi_n}\right). \tag{2.17}$$

对于定义在抽象域 $\Omega$ 上的 RKHS $H$，若 $H$ 存在框架 $F=\{\psi_n\}_{n=1}^{\infty}$，其对偶框架为 $\widetilde{F}=\{\widetilde{\psi_n}\}_{n=1}^{\infty}$，则 $H$ 的再生核可以构造为[382]

$$k(s,t)=\sum_{n=1}^{\infty}\widetilde{\psi_n}(s)\psi_n(t),\qquad \forall s,t\in\Omega. \tag{2.18}$$

### 2.4.2　希尔伯特空间中线性算子的广义逆

希尔伯特空间中线性算子的广义逆（generalized inverse）或伪逆（pseudo inverse）理论，作为矩阵之穆尔-彭罗斯广义逆（简称广义逆或伪逆）[284-286]理论的推广，始于作为穆尔学生的曾远荣先生[383]，随后迅速发展[384-386]。该理论作为希尔伯特空间中逆问题求解的基本工具，在后续章节中将被反复、直接运用，因此本小节对其定义、典型性质、逆问题求解特点和空间分解特点等进行简要介绍。

1. 线性算子广义逆的定义及典型性质

**定义 2.23（线性算子广义逆）** 设 $\mathcal{A}$ 是从希尔伯特空间 $H_1$ 到 $H_2$ 上的有界线性算子，则 $\mathcal{A}$ 之从 $H_2$ 到 $H_1$ 的广义逆定义为同时满足以下关于 $\mathcal{X}$ 的 4 个算子方程之唯一解，记为 $\mathcal{A}^{+}$[386]：

（1）$\mathcal{AXA}=\mathcal{A}$；

（2）$\mathcal{XAX}=\mathcal{X}$；

（3）$(\mathcal{AX})^{*}=\mathcal{AX}$；

（4）$(\mathcal{XA})^{*}=\mathcal{XA}$。

其中，将只满足方程（1）的 $\mathcal{X}$ 之全体称为 $\mathcal{A}$ 的$\{1\}$逆，记为 $\mathcal{A}^{\{1\}}$（其任意个体记为 $\mathcal{A}^{-}$）。类似地，可以定义 $\mathcal{A}$ 的$\{1，2\}$逆 $\mathcal{A}^{\{1,2\}}$、$\mathcal{A}$ 的$\{1，2，3\}$逆 $\mathcal{A}^{\{1,2,3\}}$。所以，$\mathcal{A}^{+}$就是 $\mathcal{A}^{\{1,2,3,4\}}$，这是唯一的。

这一定义是矩阵广义逆定义[284-286]的直接推广，其中的 4 个方程也能从逆问题求解（由算子的像求解原像）中寻找最佳逼近解时由必备条件得到[384]。

希尔伯特空间中线性算子的广义逆有如下典型性质[387-390]：

（1）自反性：$(\mathcal{A}^{+})^{+}=\mathcal{A}$；

（2）伴随互换性：$(\mathcal{A}^{+})^{*}=(\mathcal{A}^{*})^{+}$；

（3）与伴随算子的关系：$\mathcal{A}^{+}=(\mathcal{A}^{*}\mathcal{A})^{+}\mathcal{A}^{*}$，$\mathcal{A}^{+}=\mathcal{A}^{*}(\mathcal{A}\mathcal{A}^{*})^{+}$；

（4）若 $R_{\mathcal{A}}$ 为闭集，则：$(\mathcal{A}^{*}\mathcal{A})^{+}=\mathcal{A}^{+}(\mathcal{A}^{*})^{+}$，$(\mathcal{A}\mathcal{A}^{*})^{+}=(\mathcal{A}^{*})^{+}\mathcal{A}^{+}$。

2. 线性算子广义逆的逆问题求解特点

对给定的从希尔伯特空间 $H_1$ 到 $H_2$ 上的有界线性算子 $\mathcal{A}$ 和 $g\in R(\mathcal{A})\subset H_2$，由方程

$$\mathcal{A}f_0=g \tag{2.19}$$

求解 $f_0\in H_1$ 的问题称为逆问题。式（2.19）存在唯一的最优解[387, 388]，即

$$f_{\mathrm{opt}}=\mathcal{A}^{+}g,\quad \|f_{\mathrm{opt}}\|\leqslant\|f_0\|,\quad \forall f_0\in H_1. \tag{2.20}$$

注解的“最优”指范数最小。

3. 线性算子广义逆的空间分解特点

首先，对给定的从希尔伯特空间 $H_1$ 到 $H_2$ 的有界线性算子 $\mathcal{A}$，满足定义 2.23 之方程（1）的算子 $\mathcal{X}$（称为 $\mathcal{A}$ 的内逆，即 $\mathcal{A}^{\{1\}}$）给出 $H_1$ 和 $H_2$ 的一种特定分解[390]：

$$H_1=N(\mathcal{A})\dot{+}R(\mathcal{X}\mathcal{A}),\quad H_2=R(\mathcal{A})\dot{+}N(\mathcal{A}\mathcal{X}). \tag{2.21}$$

这样的 $\mathcal{X}$ 不是唯一的（不同的 $\mathcal{X}$ 给出 $H_1$ 和 $H_2$ 不同的分解，反之亦然）。

类似地，满足定义 2.23 之方程（2）的算子 $\mathcal{X}$（称为 $\mathcal{A}$ 的外逆，即 $\mathcal{A}^{\{2\}}$）给出 $H_1$ 和 $H_2$ 的一种特定分解[390]：

$$H_1=R(\mathcal{X})\dot{+}N(\mathcal{X}\mathcal{A}),\quad H_2=N(\mathcal{X})\dot{+}R(\mathcal{T}\mathcal{A}). \tag{2.22}$$

这样的 $\mathcal{X}$ 也不是唯一的（不同的 $\mathcal{X}$ 给出 $H_1$ 和 $H_2$ 不同的分解，反之亦然）。

进一步，同时满足定义 2.23 之方程（1）和方程（2）的算子 $\mathcal{X}$（即 $\mathcal{A}^{\{1,2\}}$）给出 $H_1$ 和 $H_2$ 的一种特定分解[390]：

$$H_1=N(\mathcal{A})\dot{+}R(\mathcal{X}),\quad H_2=R(\mathcal{A})\dot{+}N(\mathcal{X}). \tag{2.23}$$

这样的 $\mathcal{X}$ 称为 $\mathcal{A}$ 的代数广义逆（即 $\mathcal{A}^{\{1,2\}}$），但这样的 $\mathcal{X}$ 还不是唯一的。

最后，由投影算子的定义（定义 2.17）可知，$\mathcal{A}$ 的内逆 $\mathcal{X}$ 为 $H_1$ 和 $H_2$ 分别确定了特定的投影算子 $\mathcal{P}$ 和 $\mathcal{Q}$，使得其值域分别为 $R(\mathcal{P})=N(\mathcal{A})$ 和 $R(\mathcal{Q})=R(\mathcal{A})$，而值域的补空间均不确定。但若 $\mathcal{X}$ 同时还是 $\mathcal{A}$ 的外逆，且对于上述投影算子 $\mathcal{P}$ 和 $\mathcal{Q}$ 有

$$\mathcal{A}\mathcal{X}=\mathcal{Q},\quad \mathcal{X}\mathcal{A}=I-\mathcal{P}, \tag{2.24}$$

则这时的 $\mathcal{X}$ 是唯一的而且就是 $\mathcal{A}^{+}$，而式（2.24）中两式分别为对应于定义 2.23 的方程（3）和方程（4），且其对 $H_1$ 和 $H_2$ 的分解也是唯一的——正交分解[387, 390]：

$$H_1=N(\mathcal{A})\oplus R(\mathcal{A}^{+}),\quad H_2=R(\mathcal{A})\oplus N(\mathcal{A}^{+}), \tag{2.25}$$

这时 $\mathcal{P}$ 和 $\mathcal{Q}$ 是自伴算子（因而进一步成为正投影算子），且

$$\mathcal{P}=\mathcal{P}_{N(\mathcal{A})}=I-\mathcal{A}^{+}\mathcal{A}=I-\mathcal{P}_{R(\mathcal{A}^{+})}=I-\mathcal{P}_{R(\mathcal{A}^{*})},\quad \mathcal{Q}=\mathcal{P}_{R(\mathcal{A})}=\mathcal{A}\mathcal{A}^{+}. \tag{2.26}$$

### 2.4.3　框架与算子广义逆的关系

希尔伯特空间中的框架与算子广义逆之间有着重要的关系[387, 391]，本小节对其中的典型关系进行简单介绍。

1. 框架算子的空间分解性与框架存在的充要条件

对于希尔伯特空间 $H$ 中的框架 $F=\{\psi_n\}_{n=1}^{\infty}$，由于由式（2.9）所定义的框架算子 $\mathcal{S}$ 是有界、可逆的自伴算子，所以其逆算子 $\mathcal{S}^{-1}$ 自然满足定义 2.23 中的 4 个方程。因此，由式（2.25）可知，$\mathcal{S}$ 与 $\mathcal{S}^{-1}$ 将 $H$ 划分为

$$H=N(\mathcal{S})\oplus R(\mathcal{S}^{-1})=R(\mathcal{S})\oplus N(\mathcal{S}^{-1}). \tag{2.27}$$

另外，由于 $\mathcal{S}$ 与 $\mathcal{S}^{-1}$ 都是自伴算子，所以对任意 $f\in H$ 都有

$$f=\mathcal{S}\mathcal{S}^{-1}f=\sum_{n=1}^{\infty}<\mathcal{S}^{-1}f,\psi_n>\psi_n=\sum_{n=1}^{\infty}<f,\mathcal{S}^{-1}\psi_n>\psi_n, \tag{2.28}$$

所以，如下定理成立[387, 391]。

**定理 2.4（框架存在的充要条件）**

（1）希尔伯特空间 $H$ 中的点列 $F=\{\psi_n\}_{n=1}^{\infty}$ 成为框架的充要条件：由定义 2.21 所明确定义的算子 $\mathcal{S}$（或 $\mathcal{A}^{*}$）是满射算子。

（2）希尔伯特空间 $H$ 中的贝塞尔序列 $F=\{\psi_n\}_{n=1}^{\infty}$ 成为框架的充要条件：由定义 2.21 所明确定义的算子 $\mathcal{A}^{*}$是满射闭算子。

2. 两种框架算子之逆与广义逆的关系

在定义 2.21 中，对 $H$ 的框架 $F=\{\psi_n\}_{n=1}^{\infty}$（同时也是其闭子空间 $S$ 的框架）、对偶框架 $\widetilde{F}=\{\widetilde{\psi_n}\}_{n=1}^{\infty}$、框架算子 $\mathcal{S}$，以及任意 $f\in H$，由重构定理（定理 2.3）之式（2.16）可知：

$$\mathcal{P}_S f=\sum_{N=1}^{\infty}<\mathcal{P}_S f,\widetilde{\psi}_n>\psi_n=\sum_{n=1}^{\infty}<f,\mathcal{P}_S\mathcal{S}^{-1}\widetilde{\psi}_n>\psi_n=\sum_{n=1}^{\infty}<f,\mathcal{S}^{-1}\widetilde{\psi}_n>\psi_n. \tag{2.29}$$

所以，框架算子 $\mathcal{S}$ 之逆算子 $\mathcal{S}^{-1}$ 与框架算子 $\mathcal{A}$ 之广义逆 $\mathcal{A}^{+}$之间有如下关系[387]：

$$([\mathcal{A}^{*}]^{+}f)_n=<f,\mathcal{S}^{-1}\psi_n>,\quad \forall f\in H. \tag{2.30}$$

对比式（2.29）和式（2.30）可知，框架算子 $\mathcal{S}$ 将框架所张成子空间 $S$ 上的正投影以及式（2.19）所对应逆问题（那里的 $H_1$ 和 $H_2$ 对应于上述讨论中的 $H$ 和 $l^2$）紧

密地联系在一起，为逆问题求解计算提供了一种有效的手段（由框架及框架算子可计算原像在框架上的投影系数，进而得到原像的最佳逼近结果）。

## 2.5 本 章 小 结

代数与泛函是投影学习的理论基础，本章在简要介绍希尔伯特空间有关概念的基础上，重点对投影定理、线性算子与投影算子、框架理论和线性算子的广义逆等内容进行了集中讨论，为后续投影学习算法设计奠定了必要的基础。

# 第 3 章　描述型投影学习

## 3.1　引　　言

描述型投影学习是通过学习来提取数据的代表性特征（representative features），如统计相关性特征、拓扑结构性特征等。或者，建立代表性特征的描述性模型（descriptive model），如预测预报模型、身份表征模型等。前者通常用于发掘数据内在的结构信息，是被学界称为“表示学习（representation learning）”或“特征学习（feature learning）”中的一部分[258, 392]；后者主要用于描述变量之间的映射关系，这是机器学习的传统任务（参见 1.2.3 小节）。描述型投影学习的最基本性能要求是泛化能力强。

对图 1.2 所示的机器学习基本问题，若在度量空间中讨论，则以未知规则 $f_0(\boldsymbol{x})$ 与其学习结果 $f(\boldsymbol{x})$ 之距离 $\rho(f, f_0)$ 为最小的准则，往往因不依赖于训练样本而成为最优泛化准则；若在赋范线性空间中讨论，则以误差函数 $\Delta = f_0(\boldsymbol{x}) - f(\boldsymbol{x})$ 之范数为最小的准则，往往也是最优泛化准则。在希尔伯特空间中，由于无论是 $\rho(f, f_0)$ 还是 $\Delta$ 之范数都只能由内积（投影）确定且通常为二范数，所以投影准则是最基本的最优泛化准则。

本章在探讨希尔伯特空间中描述型投影学习基本准则和形式的基础上，讨论其推广形式和增量形式。

## 3.2　描述型投影学习的基本准则和形式

设 $f_0(\boldsymbol{x})$ 和 $f(\boldsymbol{x})$ 是定义在抽象域 $\Omega$ 上的希尔伯特空间 $H$ 中的点（$\boldsymbol{x} \in \Omega$），则误差函数 $\Delta(\boldsymbol{x}) = f_0(\boldsymbol{x}) - f(\boldsymbol{x})$ 的范数 $\|\Delta\|$ 通常为二范数。本节以 $\|\Delta(\boldsymbol{x})\|$ 为基础，探讨描述型投影学习基本准则和形式，包括确定性问题和随机性问题中投影学习的基本准则及形式。

### 3.2.1　确定性问题中的最优泛化学习准则及投影约束解

确定性问题是指 $f_0(\boldsymbol{x})$ 为确定函数，且在学习问题之训练数据获取过程中也不含随机因素，即式（1.1）对应的观测方程中不含噪声向量 $\boldsymbol{\nu}$（无噪标签问题），这是最基本的学习问题，对应的最优泛化准则为代价函数（泛函）：

$$c(f_0,f)=c(\varDelta)\equiv\|\varDelta\|^2,\ \varDelta=f_0(\boldsymbol{x})-f(\boldsymbol{x}).\tag{3.1}$$

根据投影定理（定理 2.1），若 $f_0(\boldsymbol{x})$在 $H$ 中一个闭子空间 $S$ 上的投影为 $f_{\mathrm{opt}}(\boldsymbol{x})$，则必有

$$f_{\mathrm{opt}}(\boldsymbol{x})=\mathcal{P}_S f_0(\boldsymbol{x})=\underset{\{f(\boldsymbol{x})\in S\}}{\mathrm{argmin}}\|\varDelta\|^2,\tag{3.2}$$

反之不一定成立。换言之，对给定 $S\subset H$，若限定在 $S$ 中寻找满足式（3.2）的最优解，则这个解是唯一的，且就是 $\mathcal{P}_S f_0$。如何确定这个 $S$，正是机器学习之模型选择问题中的基函数选择问题（参见图 1.3）。

设式（1.1）对应观测方程中的观测算子 $\mathcal{A}$ 是以函数系 $F=\{\psi_n\}_{n=1}^N\subset H$ 为采样函数对应的采样算子，而标签向量 $\boldsymbol{g}$ 和标签噪声向量 $\boldsymbol{\nu}$ 所属空间为希尔伯特空间 $C^N$（即 $N$ 维向量空间），且 $C^N$ 的标准函数集为 $\{\boldsymbol{e}_n\}_{n=1}^N\subset C^N$，则

$$\mathcal{A}=\sum_{n=1}^N\left(\boldsymbol{e}_n\otimes\overline{\psi_n}\right),\tag{3.3}$$

所以，

$$\boldsymbol{f}_0=\mathcal{A}f_0=\sum_{n=1}^N<f_0(\boldsymbol{x}),\psi_n(\boldsymbol{x})>\boldsymbol{e}_n,\tag{3.4}$$

或等价地

$$\boldsymbol{f}_0=[<f_0(\boldsymbol{x}),\psi_1(\boldsymbol{x})>,<f_0(\boldsymbol{x}),\psi_2(\boldsymbol{x})>,\cdots,<f_0(\boldsymbol{x}),\psi_N(\boldsymbol{x})>]^{\mathrm{T}}.\tag{3.5}$$

因此，对于确定性问题有

$$\boldsymbol{g}=\boldsymbol{f}_0=\mathcal{A}f_0,\tag{3.6}$$

此即式（2.19）（这时 $H_2=C^N$）。所以，由式（2.20）和式（2.26）可知，$f_0$ 的最优解为

$$f_{\mathrm{opt}}(\boldsymbol{x})=\mathcal{A}^+\boldsymbol{g}=\mathcal{A}^+\mathcal{A}f_0=P_{R(\mathcal{A}^*)}f_0(\boldsymbol{x}).\tag{3.7}$$

这时，$R(\mathcal{A}^*)$是满足式（3.2）的 $H$ 之所有子空间 $S$ 中最大的一个[356, 393]，取 $S=R(\mathcal{A}^*)$ 即有 $\mathcal{P}_S=\mathcal{P}_{R(\mathcal{A}^*)}$。因此，式（3.2）的投影约束等价于限制 $\mathcal{A}$ 的逆算子 $\mathcal{X}$（称为学习子），使得

$$\mathcal{X}\mathcal{A}=\mathcal{P}_{R(\mathcal{A}^*)}\Leftrightarrow\mathcal{X}\mathcal{A}f_0=\mathcal{P}_{R(\mathcal{A}^*)}f_0,\forall f_0\in H\tag{3.8}$$

成立，这一关系称为对学习子 $\mathcal{X}$ 的投影约束。由于该投影约束是算子方程 $\mathcal{X}\mathcal{A}=C$ 在 $C=\mathcal{A}^+\mathcal{A}$ 时的特例，所以其通解为[389, 394]

$$\mathcal{X}=\mathcal{A}^++\mathcal{G}(I-\mathcal{A}\mathcal{A}^+),\tag{3.9}$$

式中，$\mathcal{G}$ 为从 $C^N$ 到 $H$ 的任意算子。换言之，式（3.7）是式（1.1）所对应逆问题的投影约束解

$$f(\boldsymbol{x})=\mathcal{X}\boldsymbol{g}\tag{3.10}$$

中取 $\mathcal{X}=\mathcal{X}_{\text{opt}}=\mathcal{A}^{+}$（特解）的结果[393]，而式（3.10）则作为式（1.1）之逆问题解（学习结果）的一般形式，称为描述子（descriptor）或表示子（representor）。

式（3.7）就是确定性问题中式（3.2）所对应的最优泛化学习准则在投影约束下的解之基本形式——无噪条件下描述型投影学习的表示子。

进一步，若 $H$ 是再生核为 $k$ 的 RKHS，则式(3.3)中 $\psi_n(\boldsymbol{x})$ 可取为 $\psi_n(\boldsymbol{x})=k(\boldsymbol{x},\boldsymbol{x}_n)$，再由式（3.9）、式（3.10）和 2.4.2 小节中算子广义逆的性质 $\mathcal{A}^{+}=\mathcal{A}^{*}(\mathcal{A}\mathcal{A}^{*})^{+}$，可得如下定理[395]。

**定理 3.1（KNR）** 无噪条件下描述型投影学习用最优泛化核非线性表示子（kernel-based nonlinear representor，KNR）表示为

$$f(\boldsymbol{x})=\sum_{n=1}^{N}a(n)k(\boldsymbol{x},\boldsymbol{x}_n),\tag{3.11}$$

其中

$$\boldsymbol{a}=[a(1),a(2),\cdots,a(N)]^{\mathrm{T}}=\boldsymbol{K}^{+}\boldsymbol{g}=\boldsymbol{K}^{+}\boldsymbol{f}_0,\tag{3.12}$$

而其中第 $i$ 行、第 $j$ 列元素为

$$[\boldsymbol{K}]_{i,j}=[\mathcal{A}^{*}\mathcal{A}]_{i,j}=<\psi_i(\boldsymbol{x}),\psi_j(\boldsymbol{x})>=<k(\boldsymbol{x}_i),k(\boldsymbol{x}_j)>,\quad i,j=1,2,\cdots,N\tag{3.13}$$

的矩阵 $\boldsymbol{K}$ 称为函数系 $F=\{\psi_n\}_{n=1}^{N}$ 对应的格拉姆（Grammer）矩阵。

显然，KNR 的基本形式和网络结构如图 3.1[211]所示，为单隐层网络。

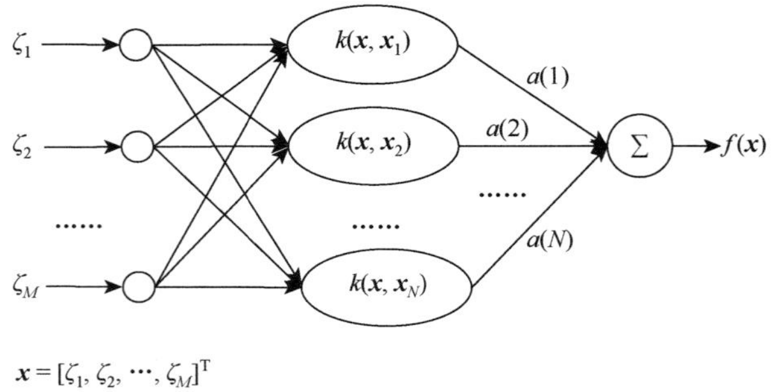


图 3.1　KNR 的基本形式和网络结构[211]

### 3.2.2　随机性问题中的最优泛化学习准则及投影约束解

随机性问题包括：$f_0(\boldsymbol{x})$为随机函数的一般情况，以及$f_0(\boldsymbol{x})$为确定函数但式（1.1）对应的训练数据获取过程中引入随机因素的特定情况。无论是何种情况，都假定

$f_0(\boldsymbol{x})$为定义在抽象域 $\Omega$ 上的希尔伯特空间 $H$（不局限于 $L^2$ 空间）中的点，且 $\boldsymbol{x}\in\Omega$。在式（3.1）之误差平方代价下对应的最优泛化准则为一般的 MSE 准则，而在 $f_0(\boldsymbol{x})$ 为确定函数、训练数据获取过程中存在随机因素的特定情况下引入投影约束后得到相应的解，就是含噪条件下的描述型投影学习形式。

1. 随机性问题中的最优泛化学习准则：MSE 准则及其特例

最小 MSE 准则的基本形式用风险表示为

$$\mathcal{R}_{\mathrm{MSE}}(f)\equiv E\left[\left\|f_0(\boldsymbol{x})-f(\boldsymbol{x})\right\|^2\right]=E\left[\left\|\Delta\right\|^2\right]=\min. \tag{3.14}$$

显然，对 $f_0(\boldsymbol{x})$为随机函数的一般情况，最小 MSE 准则有如下特例。

（1）对无噪标签问题，$g(n)=f_0(n)$，若以 $D=\{\boldsymbol{x}(n),f_0(n)\}$为经验数据并用算术平均代替数学期望则得到式（1.13），这时最小 MSE 准则就成为经典记忆学习准则——LMS 准则。

（2）对含噪标签问题，$g(n)=f_0(n)+\boldsymbol{\nu}(n)$，若以 $D=\{\boldsymbol{x}(n),g(n)\}$为经验数据并用算术平均代替数学期望则得到式（1.14），这时 MSE 准则就是 LS 准则。

总之，传统记忆学习准则是 MSE 准则经验化（近似）的结果。

（3）对无噪标签问题，$g(n)=f_0(n)$，若要求标签 $f_0(n)(n=1,2,\cdots,N)$为独立序列，则由 1.3.2 小节中的 KLT 可知，MSE 准则经验化得到的最优解 $f(\boldsymbol{x})$就是 $f_0(\boldsymbol{x})$ 在空间 $S=\mathrm{span}\{\psi_1(\boldsymbol{x}),\psi_2(\boldsymbol{x}),\cdots,\psi_N(\boldsymbol{x})\}$上的正投影，即 KLE；进一步，当原像空间 $H$ 也是有限维向量空间时，KLT 退化成在数据降维处理中得到广泛应用的 PCA，所以是典型的最优泛化正投影学习。总之，KLT 和 PCA 中的优化准则是 MSE 准则的特例。

2. MSE 准则的投影约束解：投影学习

对于式（1.1）所对应的一般逆问题，不失一般性，设 $E[\boldsymbol{\nu}]=0$。从逆问题求解的角度考虑，需要寻找 $\mathcal{A}$ 的一个逆算子（学习子）$\mathcal{X}$，使得式（3.10）中的 $f$ 成为 $f_0$ 的一个最优近似或最优描述（描述子或表示子），这正是投影学习（projection learning，PL）的基本出发点[354-357]。对于 $f_0(\boldsymbol{x})$为确定函数的情况，由式（1.1）、式（3.10）和式（3.14）可知，最小 MSE 准则简化为

$$\mathcal{R}_{\mathrm{MSE}}(\mathcal{X})=\left\|f_0-\mathcal{X}\mathcal{A}f_0\right\|^2+E\left[\left\|\mathcal{X}\boldsymbol{\nu}\right\|^2\right]=\min, \tag{3.15}$$

或等价为最优学习子

$$\mathcal{X}_{\mathrm{opt,MSE}}=\underset{\{\mathcal{X}\}}{\operatorname{argmin}}\ E\left[\left\|\mathcal{X}\boldsymbol{\nu}\right\|^2\right]\quad \text{s.t.}\quad \left\|f_0-\mathcal{X}\mathcal{A}f_0\right\|^2=\min. \tag{3.16}$$

若式（3.16）中的约束条件采用式（3.8）所对应的投影约束，则有

$$X_{\text{opt,PL}} = \underset{\{X\}}{\operatorname{argmin}} E\left[\|X\boldsymbol{v}\|^2\right] \quad \text{s.t.} \quad X\mathcal{A} = \mathcal{P}_{R(\mathcal{A}^*)}. \tag{3.17}$$

这正是最小MSE准则简化后得到的投影学习准则，其基本思想和几何意义如图3.2[211]所示，其中阴影部分为$\Delta = \Delta(\boldsymbol{x}) = \| f_0(\boldsymbol{x}) - f(\boldsymbol{x})\|$。

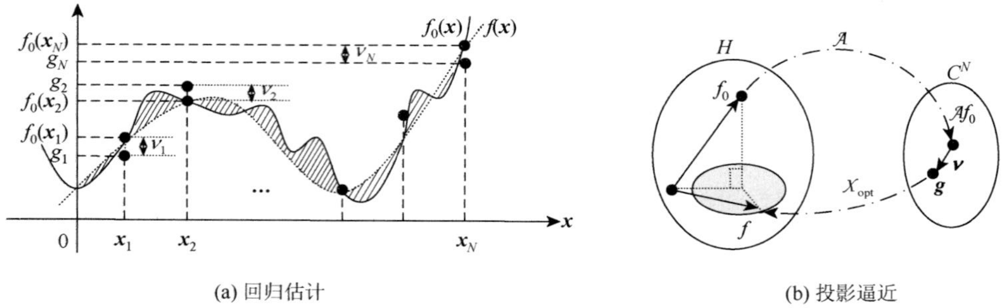


图 3.2　含噪条件下确定函数之最优泛化描述的基本思想和几何意义（投影学习）[211]

式（3.17）对应投影学习准则的通解为[354, 357]

$$X_{\text{opt,PL}} = \mathcal{V}^+ \mathcal{A}^* \boldsymbol{U}^+ + \mathcal{G}(\boldsymbol{I}_N - \boldsymbol{U}\boldsymbol{U}^+), \tag{3.18}$$

其中

$$\boldsymbol{U} = \mathcal{A}\mathcal{A}^* + \boldsymbol{Q}, \quad \mathcal{V} = \mathcal{A}^*\boldsymbol{U}^+\mathcal{A}, \tag{3.19}$$

式中，$\mathcal{G}$的含义同式（3.9）；$\boldsymbol{I}_N$为$N\times N$维单位矩阵；$\boldsymbol{Q}$为向量$\boldsymbol{v}$的相关矩阵。

若$H$是再生核为$k$的RKHS，则式（3.3）中$\psi_n(\boldsymbol{x})$可取为$\psi_n(\boldsymbol{x}) = k(\boldsymbol{x}, \boldsymbol{x}_n)$，并将式（3.10）中的$X$用式（3.18）中的$X_{\text{opt, PL}}$代入，则由式（3.3）和式（3.19）直接运算可得如下定理。

**定理 3.2（KPLR）** 基于核的投影学习表示子（kernel-based projection learning representor，KPLR）为函数

$$f_{\text{PL}}(\boldsymbol{x}) = \sum_{n=1}^{N} a_{\text{PL}}(n) k(\boldsymbol{x}, \boldsymbol{x}_n), \tag{3.20}$$

其中

$$a_{\text{PL}}(n) = \langle \boldsymbol{w}_n, \boldsymbol{h} \rangle, \ \boldsymbol{w}_n = [w_{1n}, w_{2n}, \cdots, w_{Nn})]^{\text{T}}, \ w_{mn} = [(\mathcal{V}^*\mathcal{V})^+]_{mn}[\boldsymbol{U}^+]_{mn}, \ \boldsymbol{h} = \mathcal{A}h, \tag{3.21}$$

而$m = 1, 2, \cdots, N$，且

$$h(\boldsymbol{x}) = \sum_{i=1}^{N} b(i) k(\boldsymbol{x}, \boldsymbol{x}_i), \quad \boldsymbol{b} = [b(1), b(2), \cdots, b(N)]^{\text{T}} = \boldsymbol{U}^+\boldsymbol{g}. \tag{3.22}$$

由此可知，KPLR 的映射机理和典型结构如图 3.3 所示，为级联复合结构：式（3.22）对应于噪声抑制和重建网络，即图 3.3（b）中第一隐层的下半部分（短划线框部分）；式（3.21）对应于重建再采样和权系数构建网络，即图 3.3（b）中第二隐层（点划线框部分）；式（3.20）为复合网络，即图 3.3（b）中上半部分（长划线框部分）。

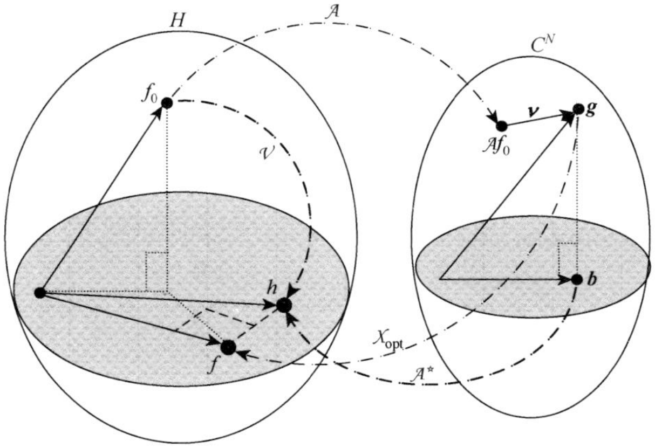


(a) 映射机理

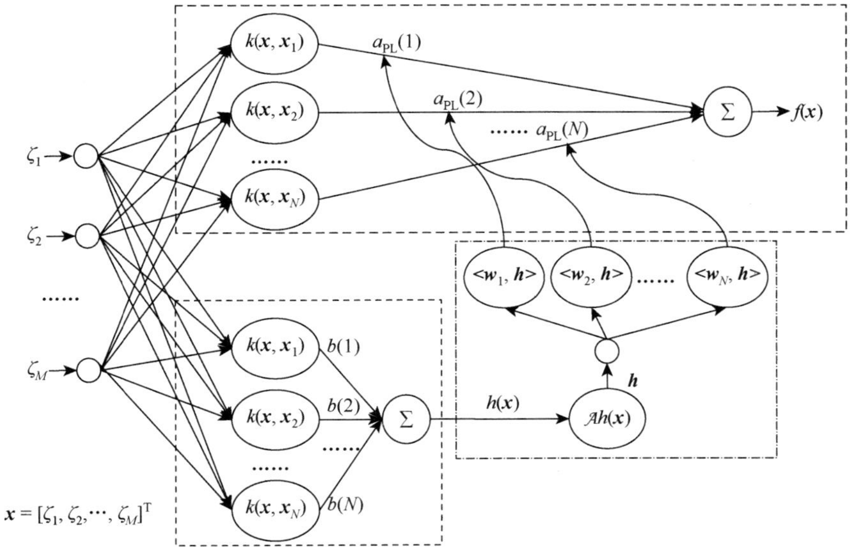


(b) 网络结构

图 3.3　KPLR 的映射机理与网络结构

另外，KPLR 也可以采用文献[356]中的形式和所对应的纯级联结构实现。

3. MSE 准则的参数化投影约束解：参数化投影学习

对于式（3.16）中的学习准则，若用参数 $\lambda(0\leqslant\lambda\leqslant1)$ 控制目标函数（噪声抑制项）和回归逼近项之间的平衡，则得到如下的参数化投影学习（parameter projection learning，PPL）准则[211, 355, 396]

$$\mathcal{X}_{\text{opt,PPL}}=\underset{\{\mathcal{X}\}}{\operatorname{argmin}}\{\lambda\operatorname{tr}[\mathcal{X}\boldsymbol{Q}\mathcal{X}^*]+(1-\lambda)\operatorname{tr}[(I-\mathcal{X}\mathcal{A})(I-\mathcal{X}\mathcal{A})^*]\},\tag{3.23}$$

式中，tr 为算子的迹。借助于文献[355]中的参数化投影准则求解引理可知，当且仅当 $\mathcal{X}_{\text{opt, PPL}}$ 满足方程

$$\mathcal{X}_{\text{opt,PPL}}[(1-\lambda)\mathcal{A}\mathcal{A}^*+\lambda\boldsymbol{Q}]=\mathcal{A}^*\tag{3.24}$$

时，$\mathcal{X}_{\text{opt, PPL}}$ 成为式（3.23）的一个解。再利用算子方程求解引理可得其一般解为[393, 394]

$$\mathcal{X}_{\text{opt,PPL}}=\mathcal{A}^*\boldsymbol{W}^++\mathcal{G}(\boldsymbol{I}_N-\boldsymbol{W}\boldsymbol{W}^+),\tag{3.25}$$

其中，$\boldsymbol{W}$ 可采用式（3.13）中的 $\boldsymbol{K}$ 和式（3.19）中的 $\boldsymbol{Q}$ 表示，即

$$\boldsymbol{W}=(1-\lambda)\boldsymbol{K}+\lambda\boldsymbol{Q}.\tag{3.26}$$

类似地，若 $H$ 为再生核为 $k$ 的 RKHS，则式（3.3）中 $\psi_n(\boldsymbol{x})$ 可取为 $\psi_n(\boldsymbol{x})=k(\boldsymbol{x},\boldsymbol{x}_n)$，并将式（3.10）中的 $\mathcal{X}$ 用式（3.25）中的学习子 $\mathcal{X}_{\text{opt, PPL}}$ 代入，得到如下定理[211, 396]。

**定理 3.3（KPPLR）** 基于核的参数化投影学习表示子（kernel-based parametric projection learning representor，KPPLR）为函数

$$f_{\text{PPL}}(\boldsymbol{x})=\sum_{n=1}^{N}a_{\text{PPL}}(n)k(\boldsymbol{x},\boldsymbol{x}_n),\tag{3.27}$$

其中，系数向量为

$$\boldsymbol{a}_{\text{PPL}}=[a_{\text{PPL}}(1),a_{\text{PPL}}(2),\cdots,a_{\text{PPL}}(N)]^{\mathrm{T}}=\boldsymbol{W}^+\boldsymbol{g}.\tag{3.28}$$

对比式（3.26）和式（3.12）可知，KNR 是当 $\lambda=0$（不存在或忽略观测噪声影响）时 KPPLR 的特例，二者有如图 3.1 所示的相同结构。

4. MSE 准则的一般解：平均投影学习准则

对于 $f_0(\boldsymbol{x})$ 为随机函数的一般情况，式（3.14）的最小 MSE 准则等价于

$$E_{f_0,f}[\operatorname{Re}\{<f_0(\boldsymbol{x}),f(\boldsymbol{x})>\}]=\max,\tag{3.29}$$

式中，Re 表示取复数的实部。可见，最小化 MSE 风险等价于最大化 $f_0(\boldsymbol{x})$ 与 $f(\boldsymbol{x})$ 的互相关。进一步，由于求互相关的内积运算为投影，所以式（3.29）等价于最大化平均投影，因此 MSE 准则又被称为平均投影学习（averaged projection learning，APL）准则[397]。由于式（3.29）中 $f_0(\boldsymbol{x})$ 和 $f(\boldsymbol{x})$ 均为未知函数，无法直接求解，只能根据机器学习问题需要（参见 1.2.3 小节），结合逆问题求解特点求解。

事实上，假设式（1.1）中的观测噪声与 $f_0(\boldsymbol{x})$不相关，则由式（1.1）和式（3.10）可知，式（3.14）的最小 MSE 准则简化为

$$\mathcal{R}_{\mathrm{APL}}(\mathcal{X})=E_{f_0}\left[\left\|f_0-\mathcal{X}\mathcal{A}f_0\right\|^2\right]+E_n\left[\left\|\mathcal{X}\boldsymbol{v}\right\|^2\right]=\min, \tag{3.30}$$

或等价为

$$\mathcal{X}_{\mathrm{opt,APL}}=\underset{\{\mathcal{X}\}}{\operatorname{argmin}}\, E_n\left[\left\|\mathcal{X}\boldsymbol{v}\right\|^2\right] \quad \text{s.t.} \quad E_{f_0}\left[\left\|f_0-\mathcal{X}\mathcal{A}f_0\right\|^2\right]=\min, \tag{3.31}$$

以此作为 APL 准则的基本形式。该准则有多种形式的解[393, 397-399]，其中形式与式（3.18）相近的学习子为[397]

$$\mathcal{X}_{\mathrm{opt,APL}}=\mathcal{R}^{1/2}\mathcal{V}_{\mathrm{APL}}^{+}\mathcal{R}^{1/2}\mathcal{A}^{*}\boldsymbol{U}_{\mathrm{APL}}^{+}+\mathcal{G}\left(\boldsymbol{I}_N-\boldsymbol{U}_{\mathrm{APL}}\boldsymbol{U}_{\mathrm{APL}}^{+}\right), \tag{3.32}$$

这里

$$\boldsymbol{U}_{\mathrm{APL}}=\mathcal{A}\mathcal{R}\mathcal{A}^{*}+\boldsymbol{Q},\quad \mathcal{V}_{\mathrm{APL}}=\mathcal{R}^{1/2}\mathcal{A}^{*}\boldsymbol{U}_{\mathrm{APL}}^{+}\mathcal{A}\mathcal{R}^{1/2}, \tag{3.33}$$

式中，$\mathcal{R}$为 $f_0(\boldsymbol{x})$的相关函数，可用诺伊曼-沙滕积表示为

$$\mathcal{R}=E_{f_0}\left[(f_0\otimes\overline{f_0})\right]. \tag{3.34}$$

类似于定理 3.2，可以得到基于核的最优泛化平均投影学习表示子（kernel-based averaged projection learning representor，KAPLR）的基本形式和网络结构，参见式（3.20）、式（3.21）、式（3.22）和图 3.3（b）。

事实上，式（3.32）是维纳滤波器的推广形式[393]。

## 3.3　描述型投影学习的扩展形式

本节所讨论的描述型投影学习的扩展形式，是上一节中多种基本形式的一般化推广和统一形式，包括偏投影学习、S-L 投影学习和偏斜投影学习。

### 3.3.1　偏投影学习

在式（3.8）所对应投影约束中，解空间 $R(\mathcal{A}^*)$是 $H$ 的所有子空间中最大的一个。通常情况下，对 $R(\mathcal{A}^*)$的任意子空间 $S\subseteq R(\mathcal{A}^*)\subset H$，任意 $f_0(\boldsymbol{x})\in H$ 在 $S$ 上的投影 $\mathcal{P}_S f_0$ 都可以作为式（3.6）所对应逆问题中满足式（3.2）所示之误差平方最小的最小范数解。若在机器学习之模型选择中确定了这个 $S$ 作为解空间，则因为 $S$ 只是 $R(\mathcal{A}^*)$的一部分，所以对应的投影学习称为部分投影或偏投影学习（partial projection learning，PTPL）[400-403]。

在此条件下，式（3.8）对应的投影约束成为偏投影约束，即[400-403]

$$\mathcal{X}\mathcal{A}\mathcal{P}_S=\mathcal{P}_{R(\mathcal{P}_S\mathcal{A}^*)}\Leftrightarrow \mathcal{X}\mathcal{A}\mathcal{P}_S f_0=\mathcal{P}_{R(\mathcal{P}_S\mathcal{A}^*)}f_0,\ \forall f_0\in H. \tag{3.35}$$

而 $S = R(\mathcal{A}^*)$ 仅作为一个特例。

在式（3.35）对应的偏投影约束下，3.2 节中多种描述型投影学习得以扩展，此处对其中的典型扩展讨论如下。

1. 偏投影学习

在偏投影约束下，式(3.17)对应的投影学习准则扩展为偏投影学习准则[400-403]

$$X_{\text{opt,PTPL}} = \underset{\{X\}}{\operatorname{argmin}} E\left[\|X\boldsymbol{v}\|^2\right] \quad \text{s.t.} \quad X\mathcal{A}\mathcal{P}_S = \mathcal{P}_{R(\mathcal{P}_S\mathcal{A}^*)}, \tag{3.36}$$

其通解的典型形式如式（3.18）和式（3.19）所示，但其中 $\boldsymbol{U}$ 和 $\mathcal{V}$ 分别替换为[400-403]

$$\boldsymbol{U}_{\text{PTPL}} = \mathcal{A}\mathcal{P}_S\mathcal{A}^* + \boldsymbol{Q}, \quad \mathcal{V}_{\text{PTPL}} = \mathcal{P}_S\mathcal{A}^*\boldsymbol{U}_{\text{PTPL}}^{+}\mathcal{A}\mathcal{P}_S. \tag{3.37}$$

由此可得对应的表示子，其被称为基于核的最优泛化偏投影学习表示子（kernel-based PTPL representor，KPTPLR），其形式和结构与 KPLR 相同（参见定理 3.2 和图 3.3），而参数不同。将式（3.19）中的 $\boldsymbol{U}$ 和 $\mathcal{V}$ 分别替换为式（3.37）中的 $\boldsymbol{U}_{\text{PTPL}}$ 和 $\mathcal{V}_{\text{PTPL}}$ 后，对应向量 $\boldsymbol{w}_n$、函数 $h(\boldsymbol{x})$ 和系数向量 $\boldsymbol{b}$ 依次被表示成 $\boldsymbol{w}_{\text{PTPL},n}$、$h_{\text{PTPL}}(\boldsymbol{x})$ 和 $\boldsymbol{b}_{\text{PTPL}}$，最后得到表示子的系数为

$$a_{\text{PTPL}}(n) = \langle \boldsymbol{w}_{\text{PTPL},n}, \boldsymbol{h}_{\text{PTPL}} \rangle, \tag{3.38}$$

式中，$\boldsymbol{h}_{\text{PTPL}} = \mathcal{A}\mathcal{P}_S\, h_{\text{PTPL}}(\boldsymbol{x})$。显然，从该表示子的形式上也能看出，KPLR 是 KPTPLR 在 $S = R(\mathcal{A}^*)$ 时的特殊情况（此时 $\mathcal{A}\mathcal{P}_S = \mathcal{A}\mathcal{P}_{R(\mathcal{A}^*)} = \mathcal{A}\mathcal{A}^+\mathcal{A} = \mathcal{A}$）。

2. 参数化偏投影学习

在偏投影约束下，式（3.23）中的参数化投影学习准则可扩展为参数化偏投影学习（parametric PTPL，PPTPL）准则：

$$X_{\text{opt,PPTPL}} = \underset{\{X\}}{\operatorname{argmin}}\{\lambda \operatorname{tr}[X\boldsymbol{Q}X^*] + (1-\lambda)\operatorname{tr}[(I - X\mathcal{A}\mathcal{P}_S)(I - X\mathcal{A}\mathcal{P}_S)^*]\}, \tag{3.39}$$

其通解的典型形式如式（3.25）所示，但其中 $\boldsymbol{W}$ 替换为

$$\boldsymbol{W}_{\text{PPTPL}} = (1-\lambda)\mathcal{A}\mathcal{P}_S\mathcal{A}^* + \lambda\boldsymbol{Q}. \tag{3.40}$$

由此得到对应的表示子称为基于核的参数化偏投影学习表示子（kernel-based PPTPL representor，KPPTPLR），其形式和结构与 KPPLR 相同（参见定理 3.3 和图 3.1）。

显然，当 $\lambda = 0$（不存在或忽略观测噪声影响）时 KPPTPLR 自然得到 KNR（定理 3.1）的扩展形式，称为核非线性偏表示子（kernel-based nonlinear partial representor，KNPTR）。

### 3.3.2　S-L 投影学习

S-L 投影学习是在平均投影学习和偏投影学习的基础上，将投影约束进一步扩展而得到的一簇投影学习，称为投影学习族[404, 405]。

1. S-L 投影学习准则

各种投影学习都有式（3.16）中的目标函数，仅有约束条件的差别：式（1.1）中的观测算子 $\mathcal{A}$ 和式（3.10）中的重构算子（学习子）$\mathcal{X}$ 之级联 $\mathcal{X}\mathcal{A}$ 成为 $f_0(\boldsymbol{x})$ 所属希尔伯特空间 $H$ 之不同子空间上的投影算子。例如，对偏投影学习而言，令

$$\mathcal{A}_S = \mathcal{A}\mathcal{P}_S, \tag{3.41}$$

则 $\mathcal{A}_S$ 为 $S \subseteq H$ 中的采样算子，对应于式（3.3）之 $H$ 中的采样算子。这时有[401]

$$S = R\left(\mathcal{A}_S^*\right) \oplus [S \cap N(\mathcal{A})], \tag{3.42}$$

于是，式（3.35）对应的偏投影约束等价于[404]

$$\mathcal{X}\mathcal{A}\mathcal{P}_S = \mathcal{P}_{[S \cap N(\mathcal{A})]^\perp}\mathcal{P}_S. \tag{3.43}$$

再如，对平均投影学习而言，限定求解空间为 $R(\mathcal{R}\mathcal{A}^*) \subseteq S \subset H$，则式（3.31）对应的约束等价于[399, 404]

$$\mathcal{X}\mathcal{A}\mathcal{R}\mathcal{A}^* = \mathcal{R}\mathcal{A}^* \Leftrightarrow \mathcal{X}\mathcal{A} = \mathcal{P}_{R(\mathcal{R}\mathcal{A}^*), \overline{\mathcal{R}} \cap N(\mathcal{A})}, \tag{3.44}$$

式中，$\mathcal{R}$ 如式（3.34）所示；$\overline{\mathcal{R}}$ 为 $\mathcal{R}$ 的闭包；$\mathcal{P}_{R(\mathcal{R}\mathcal{A}^*), \overline{\mathcal{R}} \cap N(\mathcal{A})}$ 为沿 $\overline{\mathcal{R}} \cap N(\mathcal{A})$ 向 $R(\mathcal{R}\mathcal{A}^*)$ 的斜投影。

若进一步，对子空间 $S \subseteq H$ 而言，记 $S \cap N(\mathcal{A})$ 的相对于 $S$ 的补空间为 $L$，有

$$S = L \dot{+} [S \cap N(\mathcal{A})], \tag{3.45}$$

并将 $S$ 中沿 $S \cap N(\mathcal{A})$ 向 $L$ 的斜投影算子记为 $P_{\mathrm{SL}}$，则得到 S-L 投影约束[404]：

$$\mathcal{X}\mathcal{A}\mathcal{P}_S = \mathcal{P}_{\mathrm{SL}}\mathcal{P}_S, \tag{3.46}$$

由此得到与式（3.17）对应的 S-L 投影学习准则：

$$\mathcal{X}_{\mathrm{opt,SL}} = \underset{\{\mathcal{X}\}}{\operatorname{argmin}}\, E\left[\left\|\mathcal{X}\boldsymbol{v}\right\|^2\right] \quad \text{s.t.} \quad \mathcal{X}\mathcal{A}\mathcal{P}_S = \mathcal{P}_{\mathrm{SL}}\mathcal{P}_S. \tag{3.47}$$

显然，式（3.8）中的投影约束是 S-L 投影约束在 $S = H$ 而 $L = R(\mathcal{A}^*)$ 时的特例，式（3.35）或式（3.43）中偏投影约束是 S-L 投影约束在 $S \subset H$ 而 $L = R(\mathcal{P}_S\mathcal{A}^*)$ 时的特例，而式（3.44）中的平均投影约束则是 S-L 投影约束在 $S = \overline{\mathcal{R}}$ 而 $L = R(\mathcal{R}\mathcal{A}^*)$ 时的特例[404]。所以，S-L 投影约束建立了一个学习族的框架，在其中蕴含了一簇投影学习，因此称为投影学习族[404, 405]。

2. S-L 投影学习的基本形式[404, 405]

对于式（1.1）中观测结果所属的空间 $C^N$，设其中由 $R(\mathcal{A}_S)$和 $R(\boldsymbol{Q})$构成子空间 $S_t$，则有[398, 404]

$$S_t = R(\mathcal{A}_S)\dot{+}\boldsymbol{Q}R(\mathcal{A}_S)^{\perp}, \tag{3.48}$$

由此得到 $C^N$ 的直和分解：

$$C^N = S_t \oplus S_t^{\perp} = R(\mathcal{A}_S)\dot{+}\left\{\boldsymbol{Q}R(\mathcal{A}_S)^{\perp} \oplus S_t^{\perp}\right\}. \tag{3.49}$$

进一步，记 $C^N$ 中 $S_t$ 上的投影算子为 $\mathcal{P}_{S_t}$，而沿 $\boldsymbol{Q}R(\mathcal{A}_S)^{\perp} \oplus S_t^{\perp}$ 向 $R(\mathcal{A}_S)$的斜投影算子为 $\mathcal{P}_t$的，则有如下引理[404, 405]。

**引理 3.1（S-L 投影学习）** S-L 投影学习准则中的投影学习子为

$$\mathcal{X}^{(\mathrm{SL})} = \mathcal{P}_{\mathrm{SL}}\mathcal{A}_S^{+}\mathcal{P}_t + \mathcal{G}\left(\boldsymbol{I}_N - \mathcal{P}_{S_t}\right), \tag{3.50}$$

由此得到 S-L 投影学习表示子：

$$f_{\mathrm{SL}}(\boldsymbol{x}) \equiv \mathcal{X}^{(\mathrm{SL})}\boldsymbol{g} = \mathcal{P}_{\mathrm{SL}}\mathcal{A}_S^{+}\mathcal{P}_t\boldsymbol{g}. \tag{3.51}$$

3. S-L 投影学习的等价形式[394]

为便于讨论式（3.51）所对应的 S-L 投影学习表示子的特点和结构，这里以定理的形式给出 S-L 投影学习的等价形式，并进一步给出详细证明。

1）等价性引理和定理

在限定子空间 $S \subseteq H$ 的条件下，由文献[354]之引理 3 直接得到如下引理：

**引理 3.2（学习子成为 S-L 投影学习子的充要条件）** 学习子 $\mathcal{X}$ 成为 S-L 投影学习子的充要条件是 $\mathcal{X}$ 同时满足式（3.46）和算子方程

$$\mathcal{X}\boldsymbol{Q} = \mathcal{W}\mathcal{A}_S^{*}, \tag{3.52}$$

式中，$\mathcal{W}$ 为从 $C^N$ 到 $H$ 的任意算子。

由算子方程求解特点可得到如下定理。

**定理 3.4（S-L 投影学习的等价形式）** S-L 投影学习的学习子等价于

$$\mathcal{X}^{(\mathrm{SL})} = \mathcal{P}_{\mathrm{SL}}\mathcal{P}_S\mathcal{V}_{\mathrm{SL}}^{-}\mathcal{A}_S^{*}\boldsymbol{U}_{\mathrm{SL}}^{-} + \mathcal{G}\left(\boldsymbol{I}_N - \boldsymbol{U}_{\mathrm{SL}}\boldsymbol{U}_{\mathrm{SL}}^{-}\right), \tag{3.53}$$

其中

$$\boldsymbol{U}_{\mathrm{SL}} = \mathcal{A}_S\mathcal{A}_S^{*} + \boldsymbol{Q}, \quad \mathcal{V}_{\mathrm{SL}} = \mathcal{A}_S^{*}\boldsymbol{U}_{\mathrm{SL}}^{-}\mathcal{A}_S. \tag{3.54}$$

S-L 投影学习表示子的等价形式为

$$f_{\mathrm{SL}}(\boldsymbol{x}) \equiv \mathcal{X}^{(\mathrm{SL})}\boldsymbol{g} = \mathcal{P}_{\mathrm{SL}}\mathcal{P}_S\mathcal{V}_{\mathrm{SL}}^{-}\mathcal{A}_S^{*}\boldsymbol{U}_{\mathrm{SL}}^{-}\boldsymbol{g}. \tag{3.55}$$

所以，S-L 投影学习表示子与投影学习表示子、偏投影学习表示子有相同的形式和结构（参见定理 3.2 和图 3.3），仅参数不同。

2）备用引理

定理 3.4 的证明涉及多个有关算子方程解的引理，见引理 3.3～引理 3.6。

**引理 3.3（算子方程 $\mathcal{X}\mathcal{A}=\mathcal{B}$ 有解的条件及解的一般形式）**[389, 406]对给定算子 $\mathcal{A}$ 和 $\mathcal{B}$，以下条件是等价的：

（1）关于 $\mathcal{X}$ 的算子方程 $\mathcal{X}\mathcal{A}=\mathcal{B}$ 有解；

（2）$N(\mathcal{A})\subset N(\mathcal{B})$；

（3）$\mathcal{B}\mathcal{A}^{-}\mathcal{A}=\mathcal{B}$。

这时，$\mathcal{X}\mathcal{A}=\mathcal{B}$ 的解的一般形式为

$$\mathcal{X}=\mathcal{B}\mathcal{A}^{-}+\mathcal{G}(I-\mathcal{A}\mathcal{A}^{-}),\tag{3.56}$$

式中，$\mathcal{G}$ 为任意算子；$I$ 为单位算子。

**引理 3.4（算子方程 $\mathcal{A}\mathcal{X}=\mathcal{B}$ 有解的条件及解的一般形式）**[389, 407-409]对给定算子 $\mathcal{A}$ 和 $\mathcal{B}$，以下条件是等价的：

（1）关于 $\mathcal{X}$ 的算子方程 $\mathcal{A}\mathcal{X}=\mathcal{B}$ 有解；

（2）$R(\mathcal{B})\subset R(\mathcal{A})$；

（3）$\mathcal{A}\mathcal{A}^{-}\mathcal{B}=\mathcal{B}$。

这时，$\mathcal{A}\mathcal{X}=\mathcal{B}$ 的解的一般形式为

$$\mathcal{X}=\mathcal{A}^{-}\mathcal{B}+(I-\mathcal{A}^{-}\mathcal{A})\mathcal{G},\tag{3.57}$$

式中，$\mathcal{G}$ 为任意算子；$I$ 为单位算子。显然，引理 3.4 是引理 3.3 的对偶形式[354]。

其次，由引理 3.3 和引理 3.4，可以得到如下引理[354]。

**引理 3.5（$\mathcal{A}\mathcal{B}^{-}C$ 对 $\mathcal{B}^{-}$ 的不变性）**对给定算子 $\mathcal{A}$、$\mathcal{B}$ 和 $C$，若 $N(\mathcal{A})\supset N(\mathcal{B})$ 且 $R(C)\subset R(\mathcal{A})$，则 $\mathcal{A}\mathcal{B}^{-}C$ 对在 $\mathcal{B}^{\{1\}}$ 中选择不同的 $\mathcal{B}^{-}$ 是不变的（所以通常选择为 $\mathcal{B}^{+}$）。

进一步，由于 $R(\mathcal{A})$ 的正交补就是 $N(\mathcal{A}^{*})$，若 $R(\mathcal{A})\subset R(\mathcal{B})$ 则 $N(\mathcal{A}^{*})\supset N(\mathcal{B})$，所以由引理 3.5 可得引理 3.6[354]。

**引理 3.6（$\mathcal{A}^{*}\mathcal{B}^{-}\mathcal{A}$ 对 $\mathcal{B}^{-}$ 的不变性）**对给定算子 $\mathcal{A}$ 和 $\mathcal{B}$，若 $\mathcal{B}\geqslant 0$、$R(\mathcal{A})\subset R(\mathcal{B})$，则 $\mathcal{A}^{*}\mathcal{B}^{-}\mathcal{A}\geqslant 0$ 是自伴算子，且对 $\mathcal{B}^{\{1\}}$ 中选择的不同 $\mathcal{B}$ 是不变的（所以通常选择为 $\mathcal{B}^{+}$），这时

$$N(\mathcal{A}^{*}\mathcal{B}^{-}\mathcal{A})=N(\mathcal{A}),\quad R(\mathcal{A}^{*}\mathcal{B}^{-}\mathcal{A})=R(\mathcal{A}^{*}).\tag{3.58}$$

3）定理 3.4 的证明

首先，对式（3.54）中的矩阵 $\boldsymbol{U}_{\mathrm{SL}}$，以下关系成立：

$$N(\boldsymbol{U}_{\mathrm{SL}})=N\left(\mathcal{A}_S^{*}\right)\cap N(\boldsymbol{Q}),\quad R(\boldsymbol{U}_{\mathrm{SL}})=R(\mathcal{A}_S)\dotplus R(\boldsymbol{Q}).\tag{3.59}$$

事实上，由 $\boldsymbol{U}_{\mathrm{SL}}=\mathcal{A}_S\mathcal{A}_S^{*}+\boldsymbol{Q}$ 可知 $N(\boldsymbol{U}_{\mathrm{SL}})\supset\left[N\left(\mathcal{A}_S^{*}\right)\cap N(\boldsymbol{Q})\right]$。另外，对任意 $\boldsymbol{u}\in N(\boldsymbol{U}_{\mathrm{SL}})$，由于 $\boldsymbol{Q}\geqslant 0$，而

$$<\boldsymbol{U}_{\mathrm{SL}}\boldsymbol{u},\boldsymbol{u}>=\left\|\mathcal{A}_S^{*}\boldsymbol{u}\right\|^2+<\boldsymbol{Q}\boldsymbol{u},\boldsymbol{u}>=0.\tag{3.60}$$

所以有 $\boldsymbol{Qu}=0$ 和 $\mathcal{A}_S^*\boldsymbol{u}=0$，即又有 $N(\boldsymbol{U}_{\mathrm{SL}})\subset\left[N\left(\mathcal{A}_S^*\right)\cap N(\boldsymbol{Q})\right]$。因此，有 $N(\boldsymbol{U}_{\mathrm{SL}})=N\left(\mathcal{A}_S^*\right)\cap N(\boldsymbol{Q})$，这就是式（3.59）的第一式，再取其补，同时注意到 $\boldsymbol{U}_{\mathrm{SL}}$ 和 $\boldsymbol{Q}$ 都是自伴算子（酉阵），便得到式（3.59）的第二式。

其次，对式（3.54）中的算子 $\mathcal{V}_{\mathrm{SL}}$，由引理 3.6 可知：

$$N(\mathcal{V}_{\mathrm{SL}})=N(\mathcal{A}_S),\quad R(\mathcal{V}_{\mathrm{SL}})=R\left(\mathcal{A}_S^*\right).\tag{3.61}$$

再次，对式（3.47）中 S-L 投影学习准则里的 $\mathcal{P}_{\mathrm{SL}}$、$\mathcal{P}_{\mathrm{S}}$ 和 $\mathcal{A}_S$，可以证明：

$$N(\mathcal{A}_S)\subset N(\mathcal{P}_{\mathrm{SL}}\mathcal{P}_{\mathrm{S}}),\quad R\left(\mathcal{A}_S^*\right)\supset R\left(\mathcal{P}_{\mathrm{S}}\mathcal{P}_{\mathrm{SL}}^*\right).\tag{3.62}$$

事实上，对任意 $f\in N(\mathcal{A}_S)$ 有 $\mathcal{A}_S f=\mathcal{A}\mathcal{P}_S f=0$，因此有 $\mathcal{P}_S f\in S\cap N(\mathcal{A})$，这意味着 $\mathcal{P}_{\mathrm{SL}}\mathcal{P}_S f=0$，所以式（3.62）之第一式成立，再取其正交补，同时注意到 $\mathcal{P}_S$ 是自伴算子，即得到式（3.62）的第二式。

然后，对式（3.54）中的矩阵 $\boldsymbol{U}_{\mathrm{SL}}$ 和算子 $\mathcal{V}_{\mathrm{SL}}$，由引理 3.4 可知：

$$\boldsymbol{U}_{\mathrm{SL}}^-\mathcal{A}_S=\mathcal{A}_S,\quad \mathcal{V}_{\mathrm{SL}}\mathcal{V}_{\mathrm{SL}}^-\mathcal{A}_S^*=\mathcal{A}_S^*,\quad \mathcal{V}_{\mathrm{SL}}\mathcal{V}_{\mathrm{SL}}^-\mathcal{P}_S\mathcal{P}_{\mathrm{SL}}^*=\mathcal{P}_S\mathcal{P}_{\mathrm{SL}}^*.\tag{3.63}$$

最后，由引理 3.2 和式（3.54）可得

$$\mathcal{X}\boldsymbol{U}_{\mathrm{SL}}=(\mathcal{P}_{\mathrm{SL}}\mathcal{P}_S+\mathcal{G})\mathcal{A}_S^*.\tag{3.64}$$

由于有

$$N\left(\mathcal{A}_S^*\right)\subset N\left([\mathcal{P}_{\mathrm{SL}}\mathcal{P}_S+\mathcal{G}]\mathcal{A}_S^*\right),\tag{3.65}$$

进一步由式（3.59）可知

$$N(\boldsymbol{U}_{\mathrm{SL}})\subset N\left(\mathcal{A}_S^*\right)\subset N\left([\mathcal{P}_{\mathrm{SL}}\mathcal{P}_S+\mathcal{G}]\mathcal{A}_S^*\right).\tag{3.66}$$

所以由式（3.64）、式（3.66）和引理 3.3 可知

$$\mathcal{X}=[\mathcal{P}_{\mathrm{SL}}\mathcal{P}_S+\mathcal{G}]\mathcal{A}_S^*\boldsymbol{U}_{\mathrm{SL}}^-+\mathcal{G}\left(\boldsymbol{I}_N-\boldsymbol{U}_{\mathrm{SL}}\boldsymbol{U}_{\mathrm{SL}}^-\right).\tag{3.67}$$

将式（3.67）中的 $\mathcal{X}$ 代入式（3.46），同时将式（3.54）中的 $\mathcal{V}_{\mathrm{SL}}$ 代入，并利用式（3.63）的关系，可得

$$(\mathcal{P}_{\mathrm{SL}}\mathcal{P}_S+\mathcal{G})\mathcal{V}_{\mathrm{SL}}=\mathcal{P}_{\mathrm{SL}}\mathcal{P}_S.\tag{3.68}$$

根据式（3.61）、式（3.62）和引理 3.3 可知，式（3.68）关于 $\mathcal{P}_{\mathrm{SL}}\mathcal{P}_S+\mathcal{G}$ 的解为

$$\mathcal{P}_{\mathrm{SL}}\mathcal{P}_S+\mathcal{G}=\mathcal{P}_{\mathrm{SL}}\mathcal{P}_S\mathcal{V}^-+\mathcal{W}\left(I-\mathcal{V}_{\mathrm{SL}}\mathcal{V}_{\mathrm{SL}}^-\right),\tag{3.69}$$

式中，$\mathcal{W}$ 为从 $C^N$ 到 $H$ 的任意算子。将式（3.69）代入式（3.67）并利用式（3.63），可得式（3.53）。

至此定理 3.4 得到了证明。

4）两种形式的等价性证明

即证明式（3.50）与式（3.53）是等价的。

首先，显然式（3.54）中算子 $\mathcal{V}_{\mathrm{SL}}$ 是自伴算子，而由式（3.63）可得

$$\mathcal{A}_S\left(\mathcal{V}_{\rm SL}^-\mathcal{A}_S^*\boldsymbol{U}_{\rm SL}^-\right)\mathcal{A}_S=\mathcal{A}_S\mathcal{V}_{\rm SL}^-\left(\mathcal{A}_S^*\boldsymbol{U}_{\rm SL}^-\mathcal{A}_S\right)=\mathcal{A}_S\mathcal{V}_{\rm SL}^-\mathcal{V}_{\rm SL}=\left(\mathcal{V}_{\rm SL}\mathcal{V}_{\rm SL}^-\mathcal{A}_S^*\right)^*=\mathcal{A}_S, \quad (3.70)$$

也就是说，式（3.53）中的$\mathcal{V}_{\rm SL}^-\mathcal{A}_S^*\boldsymbol{U}_{\rm SL}^-\in\mathcal{A}_S^{\{1\}}$，可用$\mathcal{A}_S^-$代替。因此，式（3.53）等价于

$$\mathcal{X}^{\rm (SL)}=\mathcal{P}_{\rm SL}\mathcal{P}_S\mathcal{A}_S^-+\mathcal{G}\left(\boldsymbol{I}_N-\boldsymbol{U}_{\rm SL}\boldsymbol{U}_{\rm SL}^-\right). \quad (3.71)$$

类似地，对引理3.1中的$\mathcal{P}_t$而言，由于$\mathcal{A}_S\mathcal{A}_S^-=\mathcal{P}_{R(\mathcal{A}_S),R(\mathcal{A}_S)^-}$，其中$R(\mathcal{A}_S)^-$是$C^N$中$R(\mathcal{A}_S)$的补空间，所以有

$$\mathcal{A}_S\left(\mathcal{A}_S^-\mathcal{P}_t\right)\mathcal{A}_S=\left(\mathcal{P}_{R(\mathcal{A}_S),R(\mathcal{A}_S)^-}\mathcal{P}_{R(\mathcal{A}_S),\boldsymbol{Q}R(\mathcal{A}_S)^\perp\oplus S_t^\perp}\right)\mathcal{A}_S=\mathcal{P}_{R(\mathcal{A}_S),R(\mathcal{A}_S)^-}\mathcal{A}_S=\mathcal{A}_S, \quad (3.72)$$

因此，$\mathcal{A}_S^-\mathcal{P}_t\in\mathcal{A}_S^{\{1\}}$，而式（3.50）中的$\mathcal{A}_S^+\mathcal{P}_t\in\mathcal{A}_S^{\{1\}}$是$\mathcal{A}_S^-$特别地取为$\mathcal{A}_S^+$的情况。也就是说，式（3.50）等价于

$$\mathcal{X}^{\rm (SL)}=\mathcal{P}_{\rm SL}\mathcal{A}_S^++\mathcal{G}\left(\boldsymbol{I}_N-\mathcal{P}_{S_t}\right). \quad (3.73)$$

其次，由于对任意正投影算子$\mathcal{P}$和任意算子$\mathcal{B}$而言，均有$\mathcal{P}(\mathcal{B}\mathcal{P})^+=(\mathcal{B}\mathcal{P})^+$[286]，所以

$$\mathcal{A}_S^+=(\mathcal{A}\mathcal{P}_S)^+=\mathcal{P}_S(\mathcal{A}\mathcal{P}_S)^+=\mathcal{P}_S\mathcal{A}_S^+, \quad (3.74)$$

因此式（3.73）等价于

$$\mathcal{X}^{\rm (SL)}=\mathcal{P}_{\rm SL}\mathcal{P}_S\mathcal{A}_S^++\mathcal{G}\left(\boldsymbol{I}_N-\mathcal{P}_{S_t}\right). \quad (3.75)$$

对比式（3.71）和式（3.75）之第一项可知，在$\mathcal{A}_S^-$特别地取为$\mathcal{A}_S^+$时，二者等价。

最后，由于式（3.50）中$\mathcal{P}_{S_t}$是$C^N$中式（3.48）对应的子空间$S_t$上的投影算子，而式（3.53）中可取$\boldsymbol{U}_{\rm SL}\boldsymbol{U}_{\rm SL}^-=\mathcal{P}_{R(\boldsymbol{U}_{\rm SL}),R(\boldsymbol{U}_{\rm SL})^-}=\mathcal{P}_{R(\boldsymbol{U}_{\rm SL}),R(\boldsymbol{U}_{\rm SL})^\perp}=\boldsymbol{U}_{\rm SL}\boldsymbol{U}_{\rm SL}^+$[354]，因此由式（3.59）可知$\mathcal{P}_{S_t}$与$\boldsymbol{U}_{\rm SL}\boldsymbol{U}_{\rm SL}^-$等价。所以式（3.71）和式（3.75）之第二项也等价。

综上，式（3.71）和式（3.75）是等价的。

4. 统一形式[394]

由上述等价形式及其证明可以得到如下定理：

**定理3.5（S-L投影学习的统一形式）** S-L投影学习之学习子的统一形式为

$$\mathcal{X}^{\rm (SL)}=\mathcal{P}_{\rm SL}\mathcal{P}_S\mathcal{V}_{\rm SL}^+\mathcal{A}_S^*\boldsymbol{U}_{\rm SL}^++\mathcal{G}\left(\boldsymbol{I}_N-\boldsymbol{U}_{\rm SL}\boldsymbol{U}_{\rm SL}^+\right), \quad (3.76)$$

其中

$$\boldsymbol{U}_{\rm SL}=\mathcal{A}_S\mathcal{A}_S^*+\boldsymbol{Q},\quad \mathcal{V}_{\rm SL}=\mathcal{A}_S^*\boldsymbol{U}_{\rm SL}^+\mathcal{A}_S. \quad (3.77)$$

因此，S-L投影学习表示子的统一形式为

$$f_{\rm SL}(\boldsymbol{x})\equiv\mathcal{X}^{\rm (SL)}\boldsymbol{g}=\mathcal{P}_{\rm SL}\mathcal{P}_S\mathcal{V}_{\rm SL}^+\mathcal{A}_S^*\boldsymbol{U}_{\rm SL}^+\boldsymbol{g}. \quad (3.78)$$

所以，S-L 投影学习表示子与投影学习表示子、偏投影学习表示子有相同的形式和结构（参见定理 3.2 和图 3.3），仅参数不同。

另外，对比式（3.37）和式（3.77），可知 $\boldsymbol{U}_{\mathrm{SL}}=\boldsymbol{U}_{\mathrm{PTPL}}$，$\mathcal{V}_{\mathrm{SL}}=\mathcal{V}_{\mathrm{PTPL}}$。因此，统一用 $\boldsymbol{U}_{\mathrm{SL}}$ 和 $\mathcal{V}_{\mathrm{SL}}$ 表示。

### 3.3.3　偏斜投影学习

在涉及动态数据问题（如流媒体处理、实时数据分析、数据流建模预测等）的机器学习中，增量学习（包括在线学习、自适应学习等）和主动学习比基于全部训练数据的批量学习具有更现实的应用价值[339-342, 358-362]。本小节在证明 S-L 投影学习族不便于增量实现的基础上，探讨一种便于增量实现的学习族——偏斜投影学习（partial oblique projection learning，PTOPL）[410]，其增量学习形式放在下一小节讨论。

1. 增量数据下 S-L 投影学习的适用范围

对给定的学习算法，在增量数据下实现增量学习和主动学习时，要求满足原算法的适用条件以保留原算法的固有优势。例如，由于 SVM 学习基于 $\varepsilon$ 不敏感绝对残差代价，其经验风险实际上是残差向量的绝对范数或 1-范数（参见 1.3.2 小节），即将优化问题的讨论限定在一般的赋范线性空间，而非特殊的希尔伯特空间，是完全依赖于历史数据的批量学习，本身不支持增量学习。在 SVM 增量学习算法的讨论中[339-342]，往往将经验风险限定为残差向量的 2-范数（即在希尔伯特空间中讨论，成为 LS-SVM[278, 336]），因此原算法的固有优势（解的稀疏性）无法保障，只能通过剪枝等手段近似完成[279, 411-414]。

在 S-L 投影学习算法中子空间 $S$ 是给定的，但在增量数据下子空间 $L$ 应保持不变，否则将成为其他学习算法。满足这一要求的必要条件由以下定理给出，其证明可以将文献[415]之定理 4.5 和定理 5.2 直接应用于式（3.45）完成。

**定理 3.6（增量数据下 S-L 投影学习的适用条件）**对给定子空间 $S$，在增量数据下 S-L 投影学习依旧适用的必要条件：特征方程

$$\mathcal{P}_S\mathcal{P}_{N(\mathcal{A})}\phi_\lambda=\lambda\phi_\lambda \tag{3.79}$$

对应于特征值 $\lambda=1$ 的重数为 1，其中 $\phi_\lambda$ 是与特征值 $\lambda$ 对应的特征函数。

需要说明的是，定理 3.6 给出的必要条件即使得到满足，在增量数据含有新息的条件下由于子空间 $S\cap N(\mathcal{A})$ 发生了变化，作为 $S$ 中 $S\cap N(\mathcal{A})$ 之补空间的 $L$ 也发生了相应变化，因此 S-L 投影学习不能继续维持。正因如此，S-L 投影学习的增量形式难以实现，为此需要引入新的投影学习族。

2. PTOPL[410]

PTOPL 是为便于实现增量学习和主动学习而设计的一种投影学习族，由于涉及斜投影算子的定义和基本性质，所以在给出 PTOPL 的定义、一般形式和基本性质前，先介绍斜投影算子的定义和性质。

1）斜投影算子的定义和性质

设 $S$ 和 $S_i(i=1,2)$均为 RKHS 空间 $H$ 的闭子空间且满足直和分解关系

$$S = S_1 \dot{+} S_2, \tag{3.80}$$

而 $\mathcal{P}_S$ 和 $\mathcal{P}_i$ 分别是 $S$ 和 $S_i$ 上的正投影算子（$i=1,2$），则对任意 $f_0 \in H$ 均有 $\mathcal{P}_S f_0 \in S$ 且可被唯一分解为

$$\mathcal{P}_S f_0 = f_1 + f_2, \quad f_i \in S_i\,(i=1,2), \tag{3.81}$$

称将 $f_0$ 映射为 $f_i$ 的唯一幂等变换 $\mathcal{P}_{S_i,S_{\{i\}}}$ 定义为 $H$ 中沿子空间 $S_{\{i\}}$ 向子空间 $S_i$ 的斜投影算子，其中

$$S_{\{i\}} = S^{\perp} \oplus S_j, \quad i,j=1,2;\, j \neq i. \tag{3.82}$$

这时 $\mathcal{P}_{S_i,S_{\{i\}}}$ 的值域和零空间分别为

$$R\left(\mathcal{P}_{S_i,S_{\{i\}}}\right) = S_i,\ N\left(\mathcal{P}_{S_i,S_{\{i\}}}\right) = S_{\{i\}}. \tag{3.83}$$

由上述定义可知

$$\mathcal{P}_S = \sum_{i=1}^{2} \mathcal{P}_{S_i,S_{\{i\}}}. \tag{3.84}$$

而以下引理给出了一种构建斜投影算子的方法（参见文献[416]和文献[417]）。

**引理 3.7（斜投影算子的构建）**斜投影算子与正投影算子有如下关系：

$$\mathcal{P}_{S_i,S_{\{i\}}} = (I - \mathcal{P}_1\mathcal{P}_2)^{-1}\mathcal{P}_1(I - \mathcal{P}_2). \tag{3.85}$$

2）PTOPL 的定义

设未知规则 $f_0(\boldsymbol{x})$所属空间为 $H$，$f_0(\boldsymbol{x})$的解空间为 $S \subseteq H$，且在式（1.1）中无观测噪声时解空间为 $S_1 \subset S$，而 $S_2$ 为 $S$ 中 $S_1$ 的补空间。对于式（3.10）中的学习子 $\mathcal{X}$，若

$$\mathcal{X}\mathcal{A} = \mathcal{P}_{S_1,S_{\{1\}}} \Leftrightarrow \mathcal{X}\mathcal{A}f(\boldsymbol{x}) = \mathcal{P}_{S_1,S_{\{1\}}} f(\boldsymbol{x}), \quad \forall f(\boldsymbol{x}) \in S \tag{3.86}$$

成立，则在投影学习框架下等价地有

$$\mathcal{X}\mathcal{A}\mathcal{P}_S f_0(\boldsymbol{x}) = \mathcal{P}_{S_1,S_{\{1\}}}\mathcal{P}_S f_0(\boldsymbol{x}), \quad \forall f_0(\boldsymbol{x}) \in H, \tag{3.87}$$

即

$$\mathcal{X}\mathcal{A}_S = \mathcal{P}_{S_1,S_{\{1\}}}\mathcal{P}_S. \tag{3.88}$$

进一步，按斜投影算子的定义、性质并由 $S_1$ 为 $S$ 的子空间可知

$$\mathcal{X}\mathcal{A}_S=\mathcal{P}_{S_1,S_{\{1\}}},\tag{3.89}$$

这就是偏斜投影约束。

需要说明的是，式（3.89）中的偏斜投影约束不同于式（3.35）中的偏投影约束：其一，这里的 $S_1$ 为 $S$ 的子空间，而在偏投影约束中没有这一限制；其二，这里只要求 $\mathcal{X}\mathcal{A}$ 成为斜投影算子即可，没有像偏投影约束那样要求 $\mathcal{X}\mathcal{A}$ 必须是正投影算子。

另外，由引理 3.3 可得到如下引理。

**引理 3.8（偏斜投影约束方程有解的条件）** 偏斜投影约束方程式（3.89）有解的充要条件是

$$N(\mathcal{A}_S)\subset S_{\{1\}}.\tag{3.90}$$

因此，可得到偏斜投影学习（PTOPL）的定义。

**定义 3.1（偏斜投影学习 PTOPL）** 满足条件

$$\mathcal{X}_{\text{opt}}=\underset{\{\mathcal{X}\}}{\operatorname{argmin}}\ E\left[\left\|\mathcal{X}\boldsymbol{v}\right\|^2\right]\quad \text{s.t.}\quad \mathcal{X}\mathcal{A}_S=\mathcal{P}_{S_1,S_{\{1\}}},\ N(\mathcal{A}_S)=N(\mathcal{A}\mathcal{P}_S)\subset S_{\{1\}}=S_2\oplus S^{\perp}\tag{3.91}$$

的学习子 $\mathcal{X}_{\text{opt}}$ 及对应表示子 $f=\mathcal{X}_{\text{opt}}\boldsymbol{g}$ 称为偏斜投影学习（PTOPL），所有满足其中约束条件的投影学习称为 PTOPL 族。

值得指出以下几点：其一，与式（3.47）对应的 S-L 投影学习准则相比，PTOPL 准则没有预先限制 $S$ 的子空间 $S_1$，所以更具一般性，且允许增量数据改变 $S_1$，便于实现增量学习；其二，式（3.89）的偏斜投影约束确保了观测无噪条件下的解空间 $S_1$，而含噪条件下学习目标函数保证解的分散程度最小，既确保了最优泛化能力又保证了拟合精度；其三，若 $f_0(\boldsymbol{x})\in H$ 但 $f_0(\boldsymbol{x})\notin S_1$，则在无噪观测条件下也不能得到准确解（得到的只是其在 $S_1$ 中的投影结果），但在含噪观测条件下解的分散程度依旧是最小的，这在一些特殊应用（如存在信道间干扰时的信号分离问题[417]）中具有现实意义。

3）PTOPL 的扩展性

由于式（3.3）中的观测算子 $\mathcal{A}$ 由观测手段（体现为 $F=\{\psi_n\}_{n=1}^N\subset H$）和观测点集 $D=\{\boldsymbol{x}(n)\}(n=1,2,\cdots,N)$ 确定（对于特定应用和给定的对应观测手段，则由观测点集确定）——客观问题，而 PTOPL 中的斜投影算子 $\mathcal{P}_{S_1,S_{\{1\}}}$ 则由模型选择确定的求解空间 $S$ 和式（3.80）对应的直和分解情况确定。在给定 $S$ 的条件下，不同的直和分解对应于不同的学习算法——主观方法。

首先，若极端地取

$$S_{\{1\}}\equiv S_2\oplus S^{\perp}=N(\mathcal{A}_S)=\left\{R\left(\mathcal{A}_S^*\right)\right\}^{\perp},\tag{3.92}$$

则有

$$S = S_2 \oplus R\left(\mathcal{A}_S^*\right), \tag{3.93}$$

进而由式（3.42）可知 $S_2 = S \cap N(\mathcal{A})$，所以这时若 $S$ 的直和分解为

$$S = S_1 \dotplus \{S \cap N(\mathcal{A})\}, \tag{3.94}$$

这正是 S-L 投影学习（对应地，$L = S_1$）。因此 PTOPL 是 S-L 投影学习的扩展。

其次，从另一角度看，对任意 $f_0(\boldsymbol{x}) \in H$，由于 $\mathcal{P}_{\text{SL}}\mathcal{P}_S f_0(\boldsymbol{x}) \in L$，所以有

$$(\mathcal{P}_{\text{SL}}\mathcal{P}_S)^2 f_0(\boldsymbol{x}) = (\mathcal{P}_{\text{SL}}\mathcal{P}_S)\{\mathcal{P}_{\text{SL}}\mathcal{P}_S f_0(\boldsymbol{x})\} = \mathcal{P}_{\text{SL}}\mathcal{P}_S f_0(\boldsymbol{x}) \Leftrightarrow (\mathcal{P}_{\text{SL}}\mathcal{P}_S)^2 = \mathcal{P}_{\text{SL}}\mathcal{P}_S.$$

进一步，由于[399]

$$R(\mathcal{P}_{\text{SL}}\mathcal{P}_S) = L = S_1, \{S \cap N(\mathcal{A})\} \oplus S^{\perp} = S_{\{1\}},$$

所以 $\mathcal{P}_{\text{SL}}\mathcal{P}_S$ 为斜投影算子。因此，S-L 投影学习是 PTOPL 在满足式（3.92）对应特定条件下取 $L = S_1$ 的特例。

另外，在 3.3.2 小节的讨论中得知，S-L 投影约束蕴含了投影约束、偏投影约束、平均投影约束，即投影学习、偏投影学习、平均投影学习均是 S-L 投影学习的特例。

综上所述，偏斜投影学习是比 S-L 投影学习更一般化的描述型投影学习，而投影学习、偏投影学习、平均投影学习、S-L 投影学习均是偏斜投影学习的特例。

4）PTOPL 的一般形式

类似于定理 3.4，PTOPL 的一般形式由以下定理给出[410]。

**定理 3.7（PTOPL 的一般形式）** PTOPL 之学习子的一般形式为

$$\mathcal{X}^{(\text{PTOPL})} = \mathcal{P}_{S_1,S_{\{1\}}}\mathcal{V}_{\text{SL}}^{+}\mathcal{A}^{*}\boldsymbol{U}_{\text{SL}}^{+} + \mathcal{G}\left(\boldsymbol{I}_N - \boldsymbol{U}_{\text{SL}}\boldsymbol{U}_{\text{SL}}^{+}\right), \tag{3.95}$$

而 PTOPL 表示子的一般形式为

$$f_{\text{PTOPL}}(\boldsymbol{x}) = \mathcal{P}_{S_1,S_{\{1\}}}\mathcal{V}_{\text{SL}}^{+}\mathcal{A}^{*}\boldsymbol{U}_{\text{SL}}^{+}\boldsymbol{g}. \tag{3.96}$$

显然，PTOPL 表示子与投影学习表示子、偏投影学习表示子、S-L 投影学习表示子有相同的形式和结构（参见定理 3.2 和图 3.3），仅参数不同。

另外，若将偏投影学习表示子记为 $f_{\text{PTPL}}(\boldsymbol{x})$，则

$$f_{\text{PTOPL}}(\boldsymbol{x}) = \mathcal{P}_{S_1,S_{\{1\}}} f_{\text{PTPL}}(\boldsymbol{x}), \quad f_{\text{PTPL}}(\boldsymbol{x}) = \mathcal{V}_{\text{SL}}^{+}\mathcal{A}^{*}\boldsymbol{U}_{\text{SL}}^{+}\boldsymbol{g}. \tag{3.97}$$

所以，PTOPL 表示子是 PTPL 表示子被进一步斜投影的结果。

5）数值计算实例[410]

**例 3.1**　考虑具有如下再生核的 RKHS 中一维带限信号的学习（重构）问题：

$$k(x,x') = \frac{\Omega}{\pi}\operatorname{sinc}\frac{\Omega}{\pi}(x - x'), \tag{3.98}$$

式中，$\Omega$ 为信号带宽；辛格函数为

$$\operatorname{sinc}(x)=\begin{cases}\dfrac{\sin(\pi x)}{\pi x}, & x\neq 0,\\ 1, & x=0.\end{cases} \tag{3.99}$$

设 $\{x_n'\}_{n=1}^{15}=\{m\pi/\Omega\}_{m=-7}^{7}$，取 $\Omega=\pi/2$、$\psi_n(x)=k(x,x_n')$。考虑 $H=\mathrm{span}\{\psi_1(x),\psi_2(x),\cdots,\psi_{15}(x)\}$、$S=\mathrm{span}\{\psi_1(x),\psi_3(x),\cdots,\psi_{15}(x)\}$、$S_1=\mathrm{span}\{\psi_5(x),\psi_7(x),\cdots,\psi_{13}(x)\}$，以及 $S_2$ 为 $S_{21}=\mathrm{span}\{\psi_1(x)+\psi_5(x),\psi_3(x)+\psi_7(x),\psi_{15}(x)+\psi_9(x)\}$ 和 $S_{22}=\mathrm{span}\{\psi_1(x)+\psi_5(x)/2,\psi_3(x)+\psi_7(x)/2,\psi_{15}(x)+\psi_9(x)/2\}$ 这两种情况。

设 $f_0(x)\in S$ 且其对应坐标为{0，0，7.89，−5.54，−0.13，−5.55，3.75，0}，即事实上 $f_0(x)\in S_1$。按式（1.1）进行观测，其中 $\mathcal{A}$ 由 $\psi_n(x)$ 按式（3.3）决定，而观测噪声是均值为 0、方差为 1 的加性高斯噪声（这时 $\boldsymbol{Q}=\boldsymbol{I}_{15}$）。

对比 PTP 学习和 PTOP 学习，图 3.4 为 $f_0(x)$（实线）及其 PTP 学习（PTPL，长划线）、$S_2=S_{21}$ 时的 PTOP 学习结果（PTOPL，$S_{21}$，方点线）和 $S_2=S_{22}$ 时的 PTOP 学习结果（PTOPL，$S_{22}$，长划加点线），其中“+”对应于观测向量 $\boldsymbol{g}$。表 3.1 则列出 $f_0(x)$ 及其对应学习结果的坐标和方差（即 $E[\|X_{\mathrm{opt}}\boldsymbol{\nu}\|^2]$）。显然，与 PTPL 相比，PTOPL 重构结果更有效。

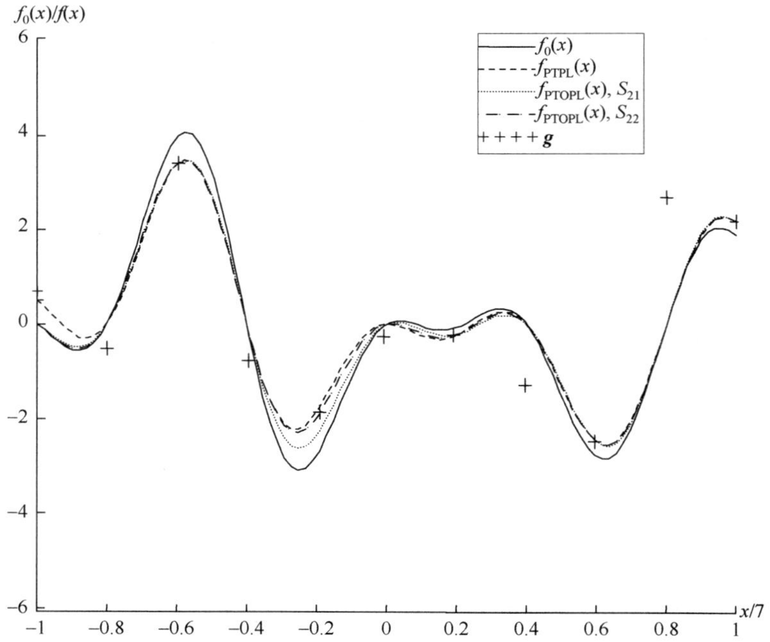


图 3.4　基于观测序列 $\boldsymbol{g}$ 对未知信号 $f_0(x)$ 的 PTPL 重构和不同 $S_2$（$S_{21}$ 和 $S_{22}$）的 PTOPL 重构

**表 3.1　未知信号 $f_0(x)$及其学习结果的坐标和方差**

| $f_0(x)$及其学习结果 | 坐标 | 方差 |
|---|---|---|
| $f_0(x)$ | {0，0，7.89，–5.54，–0.13，–5.55，3.75，0} | |
| $f_{\mathrm{PTPL}}(x)$ | {0.00，1.00，6.77，–3.69，–0.43，–4.93，4.33，0.00} | 0.13 |
| $f_{\mathrm{PTOPL}}(x), S_2 = S_{21}$ | {0.00，0.00，6.77，–4.69，–0.43，–4.93，4.33，0.00} | 0.06 |
| $f_{\mathrm{PTOPL}}(x), S_2 = S_{22}$ | {0.00，0.00，6.77，–3.94，–0.43，–4.93，4.33，0.00} | 0.11 |

## 3.4　描述型投影学习的增量形式

增量学习是利用新息对已有学习结果进行直接修正的学习算法[418-421]，在动态数据建模预测、数据流或流媒体（如音视频）信息检索、动态目标检测、实时分类等应用中具有重要意义[339-342]。对投影学习、平均投影学习而言，相应的增量学习算法也先后被设计出来[358, 359, 422]，但作为各种具体的投影学习之统一形式的 S-L 投影学习，其增量学习难以实现。本节借鉴增量投影学习，在讨论增量 PTPL 的基础上，利用 PTOPL 与 PTPL 的关系建立增量 PTOPL 算法，作为增量投影学习的统一形式[423]。

在讨论增量学习时，训练数据集是可变的，这里统一用 $n$ 表示相应的变化。例如，用 $D_n = \{\boldsymbol{x}_j, g_j\}_{j=1}^n$ 表示已有含 $n$ 对数据（样本及其标签）的数据集，而用 $\{\boldsymbol{e}_j\}_{j=1}^n$、$\mathcal{A}_n$、$\mathcal{W}_n$、$\boldsymbol{g}^{(n)}$和 $f^{(n)}(\boldsymbol{x})$分别表示 $C^n$ 的标准正交基、从 $H$ 到 $C^n$ 的映射（采样）算子、从 $C^n$ 到 $H$ 的任意算子、$n$ 个标签构成的标签向量及由 $D_n$ 得到的学习结果（表示子），等等。同时，虽然增量数据可能按多对数据（多个样本及其对应标签）的形式出现，但不失一般性，只讨论每次只增加一对数据的情况，即新增 $\{\boldsymbol{x}_{n+1}, \boldsymbol{g}_{n+1}\}$ 所得 $f^{(n+1)}(\boldsymbol{x})$与 $f^{(n)}(\boldsymbol{x})$的增量关系。

### 3.4.1　增量偏投影学习

对新增数据 $\{\boldsymbol{x}_{n+1}, \boldsymbol{g}_{n+1}\}$，由式（1.1）可知，其噪声的相关特性用 $n$ 维向量和标量分别表示为

$$\boldsymbol{q}_{n+1} = E_g[\overline{v_{n+1}}\,\boldsymbol{v}^{(n)}],\quad \sigma_{n+1} = E_g[\overline{v_{n+1}}\,v_{n+1}]. \tag{3.100}$$

若定义算子

$$\mathcal{T}_{n+1} = \sum_{j=1}^{n}\left(\boldsymbol{e}_j^{(n+1)} \otimes \overline{\boldsymbol{e}_j^{(n)}}\right), \tag{3.101}$$

则由式（3.3）可得

$$\mathcal{A}_{n+1}=\mathcal{T}_{n+1}\mathcal{A}_n+\left(\boldsymbol{e}_{n+1}^{(n+1)}\otimes\overline{\psi_{n+1}}\right),\tag{3.102}$$

式中，$\psi_{n+1}$ 为第（$n+1$）个采样函数。当 $H$ 是再生核为 $k$ 的 RKHS 时，可取 $\psi_{n+1}(\boldsymbol{x})=k(\boldsymbol{x},\boldsymbol{x}_{n+1})$。

另外，再定义 $n$ 维向量

$$\boldsymbol{s}_{n+1}=\mathcal{A}_n\mathcal{P}_S\psi_{n+1}+\boldsymbol{q}_{n+1},\tag{3.103}$$

$$\boldsymbol{t}_{n+1}=\boldsymbol{U}_{\mathrm{SL},n}^{+}\boldsymbol{s}_{n+1},\tag{3.104}$$

函数

$$\xi_{n+1}(\boldsymbol{x})=\psi_{n+1}(\boldsymbol{x})-\mathcal{A}_n^{*}\boldsymbol{t}_{n+1},\tag{3.105}$$

$$\tilde{\xi}_{n+1}(\boldsymbol{x})=\mathcal{V}_{\mathrm{SL},n}^{+}\xi_{n+1}(\boldsymbol{x}),\tag{3.106}$$

$$\tilde{\psi}_{n+1}(\boldsymbol{x})=\mathcal{P}_{N(\mathcal{A}_{S,n})}\mathcal{P}_S\psi_{n+1}(\boldsymbol{x}),\tag{3.107}$$

$$\varphi_j(\boldsymbol{x})=\mathcal{W}_n\boldsymbol{e}_j^{(n)},\ j=1,2,\cdots,n,\tag{3.108}$$

以及标量

$$\gamma_{n+1}=1+<\boldsymbol{t}_{n+1},\boldsymbol{t}_{n+1}>,\tag{3.109}$$

$$\alpha_{n+1}=<\mathcal{P}_S\psi_{n+1}(\boldsymbol{x}),\psi_{n+1}(\boldsymbol{x})>+\sigma_{n+1}-<\boldsymbol{t}_{n+1},\boldsymbol{s}_{n+1}>,\tag{3.110}$$

$$\beta_{n+1}=g_{n+1}-f_{\mathrm{PTPL}}^{(n)}(\boldsymbol{x}_{n+1})-<\boldsymbol{g}^{(n)}-\mathcal{A}_nf_{\mathrm{PTPL}}^{(n)}(\boldsymbol{x}),\boldsymbol{t}_{n+1}>,\tag{3.111}$$

式中，$f_{\mathrm{PTPL}}^{(n)}(\boldsymbol{x})$ 为由训练集 $D_n=\{\boldsymbol{x}_j,y_j\}_{j=1}^{n}$ 经 PTPL 得到的表示子。

参照文献[422]之引理 3 和引理 4，可得如下引理。

**引理 3.9**　对式（3.103）中的 $n$ 维向量 $\boldsymbol{s}_{n+1}$ 和式（3.110）中的标量 $\alpha_{n+1}$，有 $\boldsymbol{s}_{n+1}\in R(\boldsymbol{U}_{\mathrm{SL},n})$ 和 $\alpha_{n+1}\geqslant 0$。

**引理 3.10**　矩阵 $\boldsymbol{U}_{\mathrm{SL},n+1}^{+}$ 和算子 $\mathcal{V}_{\mathrm{SL},n+1}^{+}$ 的增量形式：

（1）若 $\alpha_{n+1}=0$，则

$$\boldsymbol{U}_{\mathrm{SL},n+1}^{+}=\begin{bmatrix}\mathcal{T}_{n+1}\boldsymbol{U}_{\mathrm{SL},n}^{+}\mathcal{T}_{n+1} & \mathcal{T}_{n+1}\boldsymbol{U}_{\mathrm{SL},n}^{+}\boldsymbol{t}_{n+1}/\gamma_{n+1}\\ \left(\mathcal{T}_{n+1}\boldsymbol{U}_{\mathrm{SL},n}^{+}\boldsymbol{t}_{n+1}\right)^{\mathrm{T}}/\gamma_{n+1} & <\boldsymbol{U}_{\mathrm{SL},n}^{+}\boldsymbol{t}_{n+1},\boldsymbol{t}_{n+1}>/\gamma_{n+1}^2\end{bmatrix},\tag{3.112}$$

其中，算子

$$\mathcal{T}_{n+1}=I_n-\left(\boldsymbol{t}_{n+1}\otimes\overline{\boldsymbol{t}_{n+1}}\right)/\gamma_{n+1},\tag{3.113}$$

而

$$\mathcal{V}_{\mathrm{SL},n+1}^{+}=\mathcal{V}_{\mathrm{SL},n}^{+}.\tag{3.114}$$

（2）若 $\alpha_{n+1}>0$，则

$$\boldsymbol{U}_{\mathrm{SL},n+1}^{+}=\begin{bmatrix}\boldsymbol{U}_{\mathrm{SL},n}^{+}+\left(\boldsymbol{t}_{n+1}\otimes\overline{\boldsymbol{t}_{n+1}}\right)/\alpha_{n+1} & -\boldsymbol{t}_{n+1}/\alpha_{n+1}\\ -\boldsymbol{t}_{n+1}^{\mathrm{T}}/\alpha_{n+1} & 1/\alpha_{n+1}\end{bmatrix},\tag{3.115}$$

而当 $\mathcal{P}_S\psi_{n+1} \in R(\mathcal{A}_{S,n}^*)$ 时，有

$$\mathcal{V}_{\mathrm{SL},n+1}^{+} = \mathcal{V}_{\mathrm{SL},n}^{+} - \left(\tilde{\xi}_{n+1} \otimes \overline{\tilde{\xi}_{n+1}}\right) \Big/ \left(\alpha_{n+1} + < \tilde{\xi}_{n+1}, \tilde{\xi}_{n+1} >\right), \tag{3.116}$$

当 $\mathcal{P}_S\psi_{n+1} \notin R(\mathcal{A}_{S,n}^*)$ 时，有

$$\mathcal{V}_{\mathrm{SL},n+1}^{+} = \mathcal{V}_{\mathrm{SL},n}^{+} + \frac{\alpha_{n+1} + < \tilde{\xi}_{n+1}, \tilde{\xi}_{n+1} >}{\left[\tilde{\psi}_{n+1}(\boldsymbol{x}_{n+1})\right]^2}\left(\tilde{\psi}_{n+1} \otimes \overline{\tilde{\psi}_{n+1}}\right) - \frac{\left(\tilde{\xi}_{n+1} \otimes \overline{\tilde{\psi}_{n+1}}\right) + \left(\tilde{\psi}_{n+1} \otimes \overline{\tilde{\xi}_{n+1}}\right)}{\tilde{\psi}_{n+1}(\boldsymbol{x}_{n+1})}. \tag{3.117}$$

据此得到有关增量偏投影学习的如下定理[423]，其证明类似于增量投影学习[422]。

**定理 3.8（增量 PTPL）** 从训练集 $D_{n+1} = \{\boldsymbol{x}_j, g_j\}_{j=1}^{n+1}$ 得到 PTPL 表示子的增量形式为[423]

$$f_{\mathrm{PTPL}}^{(n+1)} = \begin{cases} f_{\mathrm{PTPL}}^{(n)}, & \alpha_{n+1} = 0, \\ f_{\mathrm{PTPL}}^{(n)} + \beta_{n+1}\eta_{n+1}, & \alpha_{n+1} > 0, \end{cases} \tag{3.118}$$

其中

$$\eta_{n+1} = \begin{cases} \dfrac{\tilde{\xi}_{n+1}}{\alpha_{n+1} + < \tilde{\xi}_{n+1}, \xi_{n+1} >}, & \mathcal{P}_S\psi_{n+1} \in R(\mathcal{A}_{S,n}^*), \\ \dfrac{\tilde{\psi}_{n+1}}{\tilde{\psi}_{n+1}(\boldsymbol{x}_{n+1})}, & \mathcal{P}_S\psi_{n+1} \notin R(\mathcal{A}_{S,n}^*). \end{cases} \tag{3.119}$$

由于当 $\mathcal{P}_S = I$、$S = R(\mathcal{A}^*)$时 PTPL 就是投影学习，所以这时 PTPL 表示子的增量形式就是文献[422]中给出的投影学习表示子的增量形式。

### 3.4.2 增量 PTOPL

在机器学习中，由模型选择所确定的解空间越大，学习算法的泛化能力越强。为此，本小节根据这一思想，在对 PTOPL 进行略加限制并给出其一般形式的情况下，探讨其增量学习形式。

1）PTOPL 的限制形式——SPTOPL[423]

在由式（3.91）给出的 PTOPL 准则中，若取

$$S_{\{1\}} = S_2 \oplus S^{\perp} = N(\mathcal{A}_S) = N(\mathcal{A}\mathcal{P}_S), \tag{3.120}$$

则观测无噪条件下解空间 $S_1 = R\left(\mathcal{P}_{S_1, S_{\{1\}}}\right)$ 为 $S$ 的最大子空间，此时式（3.94）成立。由此得到 PTOPL 的限制形式——SPTOPL（special partial oblique projection learning），即有如下定理，其证明与定理 3.7 类似。

**定理 3.9（SPTOPL 的一般形式）** SPTOPL 之学习子的一般形式为[423]

$$\mathcal{X}^{(\mathrm{SPTOPL})}=\mathcal{P}_{S_1,N(\mathcal{A}_S)}\mathcal{V}_{\mathrm{SL}}^{+}\mathcal{A}^{*}\boldsymbol{U}_{\mathrm{SL}}^{+}+\mathcal{G}\left(\boldsymbol{I}_N-\boldsymbol{U}_{\mathrm{SL}}\boldsymbol{U}_{\mathrm{SL}}^{+}\right), \tag{3.121}$$

而 SPTOPL 表示子的一般形式为

$$f_{\mathrm{SPTOPL}}(\boldsymbol{x})=\mathcal{P}_{S_1,N(\mathcal{A}_S)}\mathcal{V}_{\mathrm{SL}}^{+}\mathcal{A}^{*}\boldsymbol{U}_{\mathrm{SL}}^{+}\boldsymbol{g}=\mathcal{P}_{S_1,N(\mathcal{A}_S)}f_{\mathrm{PTPL}}(\boldsymbol{x}). \tag{3.122}$$

进一步，SPTOPL 表示子与投影学习表示子、偏投影学习表示子、S-L 投影学习表示子有相同的形式和结构（参见定理 3.2 和图 3.3），仅参数不同。

2）SPTOPL 的等价形式[423]

为便于实现增量 SPTOPL，需要恰当地构建其学习子，得到对应的表示子。其中一种具体的构建方式由以下定理得到[423]。

**定理 3.10（SPTOPL 的等价形式）** SPTOPL 之学习子等价为

$$\mathcal{X}^{(\mathrm{SPTOPL})}=\left(I+\mathcal{P}_{N(\mathcal{A}_S)}\mathcal{W}\mathcal{A}_S\right)\mathcal{V}_{\mathrm{SL}}^{+}\mathcal{A}^{*}\boldsymbol{U}_{\mathrm{SL}}^{+}+\mathcal{G}\left(\boldsymbol{I}_N-\boldsymbol{U}_{\mathrm{SL}}\boldsymbol{U}_{\mathrm{SL}}^{+}\right), \tag{3.123}$$

式中，$I$ 为单位算子；$\mathcal{W}$ 为从 $C^n$ 到 $H$ 的任意算子，其他量的含义不变。对应地，SPTOPL 之表示子等价为

$$f_{\mathrm{SPTOPL}}(\boldsymbol{x})=\left(I+\mathcal{P}_{N(\mathcal{A}_S)}\mathcal{W}\mathcal{A}_S\right)\mathcal{V}_{\mathrm{SL}}^{+}\mathcal{A}^{*}\boldsymbol{U}_{\mathrm{SL}}^{+}\boldsymbol{g}=\left(I+\mathcal{P}_{N(\mathcal{A}_S)}\mathcal{W}\mathcal{A}_S\right)f_{\mathrm{PTPL}}(\boldsymbol{x}). \tag{3.124}$$

3）定理 3.10 的证明

先给出证明定理 3.10 所需的两个引理。

**引理 3.11（PTOPL 中的正、斜投影算子的特殊关系）** PTOPL 中的正投影算子 $\mathcal{P}_S$、$\mathcal{P}_2$ 与斜投影算子 $\mathcal{P}_{S_1,S_{\{1\}}}$ 有如下关系：

$$\mathcal{P}_S=(I-\mathcal{P}_2)\mathcal{P}_{S_1,S_{\{1\}}}+\mathcal{P}_2. \tag{3.125}$$

事实上，由式（3.84），以及文献[415]之定理 4.1、定理 4.2，可得

$$\mathcal{P}_{S_1,S_{\{1\}}}=(I-\mathcal{P}_1\mathcal{P}_2)^{-1}\mathcal{P}_1(I-\mathcal{P}_1\mathcal{P}_2), \tag{3.126}$$

$$\mathcal{P}_{S_2,S_{\{2\}}}=\mathcal{P}_2(I-\mathcal{P}_1\mathcal{P}_2)^{-1}(I-\mathcal{P}_1). \tag{3.127}$$

再由式（3.126）可得

$$\mathcal{P}_{S_1,S_{\{1\}}}=I-(I-\mathcal{P}_1\mathcal{P}_2)^{-1}(I-\mathcal{P}_1), \tag{3.128}$$

因此，由式（3.84）、式（3.127）和式（3.128）可得

$$\mathcal{P}_S=I-(I-\mathcal{P}_2)(I-\mathcal{P}_1\mathcal{P}_2)^{-1}(I-\mathcal{P}_1). \tag{3.129}$$

由式（3.128）和式（3.129）即可得到式（3.125）。

另外，对任意给定算子 $C$ 及其 $C^{\{1,2,3\}}$ 和 $C^{+}$，如下引理成立。

**引理 3.12（算子的{1，2，3}逆与广义逆的关系）** 对任意给定算子 $C$，$C^{\{1,2,3\}}$ 与 $C^{+}$ 有如下关系：

$$C^{\{1,2,3\}}=C^{+}+(I-C^{+}C)C_0, \tag{3.130}$$

式中，$C_0$ 为满足条件 $R\left(C_0^{*}\right)\subset R(C)$ 的任意算子。

事实上，式（3.130）中的 $C^{\{1,2,3\}}$ 满足方程 $CC^{\{1,2,3\}}C=C$ 和方程 $(CC^{\{1,2,3\}})^{*}=$

$CC^{\{1,2,3\}}$，将其代入方程 $C^{\{1,2,3\}}CC^{\{1,2,3\}}=C^{\{1,2,3\}}$ 可得 $C_0CC^{+}=C_0$，此即 $R\left(C_0^*\right)\subset R(C)$。

接下来证明定理 3.10。

显然，式（3.120）等价于式（3.92），由此得到式（3.93），因此有

$$\mathcal{P}_S=\mathcal{P}_2+\mathcal{P}_{R\left(\mathcal{A}_S^*\right)}. \tag{3.131}$$

由于 $\mathcal{P}_S\mathcal{P}_{S_1,N(\mathcal{A}_S)}=\mathcal{P}_{S_1,N(\mathcal{A}_S)}$，由引理 3.11 和式（3.131）可得

$$\mathcal{P}_{R\left(\mathcal{A}_S^*\right)}=\mathcal{P}_{R\left(\mathcal{A}_S^*\right)}\mathcal{P}_{S_1,N(\mathcal{A}_S)}. \tag{3.132}$$

再因 $\mathcal{P}_{R\left(\mathcal{A}_S^*\right)}$ 为 $R\left(\mathcal{A}_S^*\right)$ 上的正投影，所以进一步可得

$$\mathcal{P}_{R\left(\mathcal{A}_S^*\right)}=\mathcal{P}_{R\left(\mathcal{A}_S^*\right)}\mathcal{P}_{S_1,N(\mathcal{A}_S)}\mathcal{P}_{R\left(\mathcal{A}_S^*\right)}, \tag{3.133}$$

$$\left(\mathcal{P}_{R\left(\mathcal{A}_S^*\right)}\mathcal{P}_{S_1,N(\mathcal{A}_S)}\right)^*=\mathcal{P}_{R\left(\mathcal{A}_S^*\right)}\mathcal{P}_{S_1,N(\mathcal{A}_S)}. \tag{3.134}$$

另外，因 $N(\mathcal{A}_S)$ 是 $\mathcal{P}_{S_1,N(\mathcal{A}_S)}$ 的零空间而有 $\mathcal{P}_{S_1,N(\mathcal{A}_S)}\mathcal{P}_{N(\mathcal{A}_S)}=0$，继而有

$$\mathcal{P}_{S_1,N(\mathcal{A}_S)}\mathcal{P}_{R\left(\mathcal{A}_S^*\right)}=\mathcal{P}_{S_1,N(\mathcal{A}_S)}(I-\mathcal{P}_{N(\mathcal{A}_S)})=\mathcal{P}_{S_1,N(\mathcal{A}_S)}. \tag{3.135}$$

由式（3.132）和式（3.135）可得

$$\mathcal{P}_{S_1,N(\mathcal{A}_S)}=\mathcal{P}_{S_1,N(\mathcal{A}_S)}\mathcal{P}_{R\left(\mathcal{A}_S^*\right)}\mathcal{P}_{S_1,N(\mathcal{A}_S)}. \tag{3.136}$$

因此，由式（3.133）、式（3.134）和式（3.136）可知，$\mathcal{P}_{S_1,N(\mathcal{A}_S)}$ 是 $\mathcal{P}_{R\left(\mathcal{A}_S^*\right)}$ 的{1，2，3}逆。所以，注意由式（2.26）可知 $\mathcal{P}_{R\left(\mathcal{A}_S^*\right)}=\mathcal{A}_S^+\mathcal{A}_S$，因而由算子广义逆的性质可知 $\mathcal{P}_{R\left(\mathcal{A}_S^*\right)}^+=\mathcal{P}_{R\left(\mathcal{A}_S^*\right)}$ 且 $\mathcal{P}_{R\left(\mathcal{A}_S^*\right)}^+\mathcal{P}_{R\left(\mathcal{A}_S^*\right)}=\mathcal{P}_{R\left(\mathcal{A}_S^*\right)}=\mathcal{A}_S^+\mathcal{A}_S$，再由引理 3.12 和式（3.121）可得

$$\mathcal{X}^{(\mathrm{SPTOPL})}=\mathcal{P}_{R\left(\mathcal{A}_S^*\right)}\mathcal{V}_{\mathrm{SL}}^+\mathcal{A}^*\boldsymbol{U}_{\mathrm{SL}}^+ +\left(I-\mathcal{A}_S^+\mathcal{A}_S\right)C_0\mathcal{V}_{\mathrm{SL}}^+\mathcal{A}^*\boldsymbol{U}_{\mathrm{SL}}^+ +\mathcal{G}\left(\boldsymbol{I}_N-\boldsymbol{U}_{\mathrm{SL}}\boldsymbol{U}_{\mathrm{SL}}^+\right), \tag{3.137}$$

其中

$$N(C_0)\supset N(\mathcal{A}_S), \tag{3.138}$$

此式等价于[424]

$$C_0=\mathcal{W}\mathcal{A}_S=\mathcal{W}\mathcal{A}\mathcal{P}_S, \tag{3.139}$$

式中，$\mathcal{W}$ 为从 $C^n$ 到 $H$ 的任意算子。另外，类似于式（3.74），由文献[286]可知，

$$\mathcal{P}_{R\left(\mathcal{A}_S^*\right)}\mathcal{V}_{\mathrm{SL}}^+=\mathcal{V}_{\mathrm{SL}}^+, \tag{3.140}$$

因此，考虑到 $(I-\mathcal{A}_S^+\mathcal{A}_S)=\mathcal{P}_{N(\mathcal{A}_S)}$，由式（3.137）、式（3.139）和式（3.140）便得到式（3.123），进一步得到式（3.124）。定理 3.10 得证。

4）增量 SPTOP 学习[423]

引入函数

$$\lambda_{n+1} = <\mathcal{P}_S f_{\text{PTPL}}^{(n)}, \psi_{n+1}> \mathcal{P}_{N\left(\mathcal{A}_{S,n}\right)} \varphi_{n+1}, \tag{3.141}$$

其中，$\varphi_{n+1}$ 为式（3.108）中当 $j = n + 1$ 时 $\varphi_j$ 的情形。再引入算子

$$\mathcal{A}_{n+1} = I + \mathcal{P}_{N\left(\mathcal{A}_{S,n}\right)}\left(\mathcal{W}_n \mathcal{A}_{S,n} + \varphi_{n+1} \otimes \overline{\mathcal{P}_S \psi_{n+1}}\right), \tag{3.142}$$

则可得如下定理。

**定理 3.11（增量 SPTOPL）** 从训练集 $D_{n+1} = \{\boldsymbol{x}_j, g_j\}_{j=1}^{n+1}$ 得到 SPTOPL 之表示子的增量形式为[423]

$$f_{\text{SPTOPL}}^{(n+1)} = \begin{cases} f_{\text{SPTOPL}}^{(n)} + \lambda_{n+1}, & \alpha_{n+1} = 0, \\ f_{\text{SPTOPL}}^{(n)} + \beta_{n+1} \mathcal{A}_{n+1} \eta_{n+1} + \lambda_{n+1}, & \alpha_{n+1} > 0, \mathcal{P}_S \psi_{n+1} \in R\left(\mathcal{A}_{S,n}^*\right), \\ f_{\text{SPTOPL}}^{(n)} + \beta_{n+1} (\mathcal{A}_{n+1} - \mathcal{A}_0) \eta_{n+1} + \lambda_{n+1} - \lambda_0, & \alpha_{n+1} > 0, \mathcal{P}_S \psi_{n+1} \notin R\left(\mathcal{A}_{S,n}^*\right), \end{cases} \tag{3.143}$$

其中

$$\mathcal{A}_0 = \mathcal{W}_n \mathcal{A}_{S,n} + \left(\varphi_{n+1} \otimes \overline{\mathcal{P}_S \psi_{n+1}}\right), \quad \lambda_0 = \Lambda_0 f_{\text{PTPL}}^{(n)}. \tag{3.144}$$

5）定理 3.11 的证明

先给出证明定理 3.11 所需的三个引理。

**引理 3.13**　$\mathcal{W}_{n+1} \mathcal{A}_{S,n+1}$ 有如下增量形式

$$\mathcal{W}_{n+1} \mathcal{A}_{S,n+1} = \mathcal{W}_n \mathcal{A}_{S,n} + \left(\varphi_{n+1} \otimes \overline{\mathcal{P}_S \psi_{n+1}}\right). \tag{3.145}$$

事实上，由式（3.108）可得

$$\mathcal{W}_n = \sum_{j=1}^{n} \left(\varphi_j \otimes \overline{\boldsymbol{e}_j^{(n)}}\right). \tag{3.146}$$

因此，对任意 $\varphi_{n+1} \in H$，有

$$\mathcal{W}_{n+1} = \mathcal{W}_n \mathcal{T}_{n+1}^* + \left(\varphi_{n+1} \otimes \overline{\boldsymbol{e}_{n+1}^{(n+1)}}\right). \tag{3.147}$$

利用式（3.3），将式（3.101）和式（3.102）代入式（3.147）即可得式（3.145）。

**引理 3.14**　$\alpha_{n+1} = 0$ 时有

$$\mathcal{P}_S \xi_{n+1} = 0, \tag{3.148}$$

$$\mathcal{P}_S \psi_{n+1} \in R\left(\mathcal{A}_{S,n}^*\right). \tag{3.149}$$

事实上，式（3.148）可参照文献[422]之引理 5 得到，再结合式（3.105），可得式（3.149）。

**引理 3.15**　$\mathcal{A}_{S,n+1}^+ \mathcal{A}_{S,n+1}$ 有如下增量形式

$$\mathcal{A}_{S,n+1}^{+}\mathcal{A}_{S,n+1}=\begin{cases}\mathcal{A}_{S,n}^{+}\mathcal{A}_{S,n}, & \alpha_{n+1}=0,\\ \mathcal{A}_{S,n}^{+}\mathcal{A}_{S,n}, & \alpha_{n+1}>0, \mathcal{P}_S\psi_{n+1}\in R\left(\mathcal{A}_{S,n}^{*}\right),\\ I+\mathcal{A}_{S,n}^{+}\mathcal{A}_{S,n}, & \alpha_{n+1}>0, \mathcal{P}_S\psi_{n+1}\notin R\left(\mathcal{A}_{S,n}^{*}\right).\end{cases} \tag{3.150}$$

事实上，利用式（3.41），由式（3.102）可得

$$\mathcal{A}_{S,n+1}=\begin{bmatrix}\mathcal{A}_{S,n}\\ \cdots\\ (\mathcal{P}_S\psi_{n+1})^{*}\end{bmatrix}, \tag{3.151}$$

再由文献[286]之定理 4.3 和式（3.151）可得

$$\mathcal{A}_{S,n+1}^{+}=\left[\{I-\kappa_{n+1}(\mathcal{P}_S\psi_{n+1})^{*}\}\mathcal{A}_{S,n}^{+}\quad\cdots\quad\kappa_{n+1}\right], \tag{3.152}$$

其中

$$\kappa_{n+1}=\begin{cases}\dfrac{\mathcal{A}_{S,n}^{+}\left(\mathcal{A}_{S,n}^{*}\right)^{+}\mathcal{P}_S\psi_{n+1}}{1+\left\|\left(\mathcal{A}_{S,n}^{*}\right)^{+}\mathcal{P}_S\psi_{n+1}\right\|^{2}}, & \mathcal{P}_S\psi_{n+1}\in R\left(\mathcal{A}_{S,n}^{*}\right),\\ \dfrac{\left(I-\mathcal{A}_{S,n}^{+}\mathcal{A}_{S,n}\right)\mathcal{P}_S\psi_{n+1}}{\left\|\left(I-\mathcal{A}_{S,n}^{+}\mathcal{A}_{S,n}\right)\mathcal{P}_S\psi_{n+1}\right\|^{2}}, & \mathcal{P}_S\psi_{n+1}\notin R\left(\mathcal{A}_{S,n}^{*}\right).\end{cases} \tag{3.153}$$

由式（3.151）和式（3.152）可得

$$\mathcal{A}_{S,n+1}^{+}\mathcal{A}_{S,n+1}=\begin{cases}\mathcal{A}_{S,n}^{+}\mathcal{A}_{S,n}, & \mathcal{P}_S\psi_{n+1}\in R\left(\mathcal{A}_{S,n}^{*}\right),\\ I+\mathcal{A}_{S,n}^{+}\mathcal{A}_{S,n}, & \mathcal{P}_S\psi_{n+1}\notin R\left(\mathcal{A}_{S,n}^{*}\right),\end{cases} \tag{3.154}$$

再结合引理 3.14，可知式（3.150）成立。

然后再证明定理 3.11。

显然，将引理 3.13、引理 3.15 和定理 3.8 的结论代入式（3.124）即得式（3.143），所以定理 3.11 成立。

6）SPTOPL 和增量 SPTOPL 的统一性

注意到在由定理 3.10 和定理 3.11 分别给出的 SPTOPL 和增量 SPTOPL 的过程中，涉及从 $C^n$ 到 $H$ 的任意算子 $\mathcal{W}$ 或 $\mathcal{W}_n$ 的选择问题，或等价地，式（3.108）中函数系 $\{\varphi_i\}_{i=1}^{n}$ 的计算问题。这个问题产生的根本原因在于 SPTOPL 中满足式（3.120）的 $S$ 之最大子空间即式（3.80）中的 $S_1$ 有无数种选择。正因为如此，增量 SPTOPL 包含了其他投影学习的增量形式，具有统一性：对固定的 $S$ 和 $\mathcal{A}_n$，由满足不同限制条件的 $\mathcal{W}_n$ 可得到不同的投影学习及其对应的增量形式。

例如，若 $\mathcal{W}_n$ 限制为 $R(\mathcal{W}_n)\subseteq R\left(\mathcal{A}_{S,n}^*\right)$（或特别地取 $\mathcal{W}_n=\mathcal{A}_{S,n}^*$），则 SPTOPL 和增量 SPTOPL 就分别成为 PTPL 和增量 PTPL，并在 $S=H$ 和 $\mathcal{P}_S=I$ 的条件下进一步成为投影学习和增量投影学习[422]。

类似地，若 $S=\overline{R(\mathcal{R})}$ 而 $\mathcal{W}_n=\mathcal{R}\mathcal{A}^*$，则 SPTOPL 和增量 SPTOPL 就分别成为 APL 和增量 APL[357, 358, 397]。

事实上，通过对 $\mathcal{W}$ 或 $\mathcal{W}_n$ 值域的其他具体限定，就可以由 SPTOPL 和增量 SPTOPL 得到其他（新的）投影学习和对应增量学习。

7）数值计算实例[423]

**例 3.2**　考虑例 3.1 中信号复原的增量学习问题。

取 $\{x_n'\}_{n=1}^{11}=\{m\pi/\Omega\}_{m=-5}^{5}$、$\Omega=\pi/2$ 得到 $\{\psi_n(x)\}_{n=1}^{11}=\left\{k\left(x,x_n'\right)\right\}_{n=1}^{11}$，设 $H=\mathrm{span}\{\psi_1(x),\psi_2(x),\cdots,\psi_{11}(x)\}$，$S=\mathrm{span}\{\psi_1(x),\psi_3(x),\cdots,\psi_{11}(x)\}$。

设 $f_0(\boldsymbol{x})\in S$ 且其对应坐标为（−1.66，0.59，−2.67，1.43，3.25，−1.38），取采样点 $\{x_j\}_{j=1}^{7}=\{m\pi/\Omega\}_{m=-3}^{3}$、$\Omega=\pi/2$。按式（1.1）进行观测，其中 $\mathcal{A}$ 由式（3.3）决定，而观测噪声是均值为 0、方差为 1 的加性噪声（这时 $\boldsymbol{Q}=\boldsymbol{I}_7$）。在 SPTOPL 中取 $\mathcal{W}=\mathcal{A}^*$。

对比增量 PTPL 和增量 SPTOPL，图 3.5（a）～（d）给出 $f_0(\boldsymbol{x})$（实线）及得到训练数据依次为一对、三对、五对、七对时增量 PTPL 结果（方点线）和增量 SPTOPL 结果（长划加点线），其中“＋”对应于观测向量 $\boldsymbol{g}$。表 3.2 则列出 $f_0(x)$ 及其对应增量学习结果的坐标和方差（即 $E[\|\mathcal{X}_{\mathrm{opt}}\boldsymbol{\nu}\|^2]$）。

由于在实验设计中偶数号训练数据对未带来新息，所以在图中没有列出，而在表中则可明显地看到。可见，增量学习为样本和模型（表示子）的稀疏化奠定了基础。

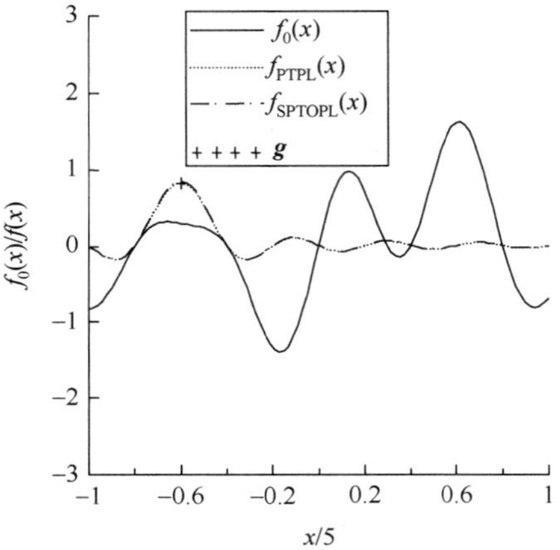

(a) 得到第一对训练数据时

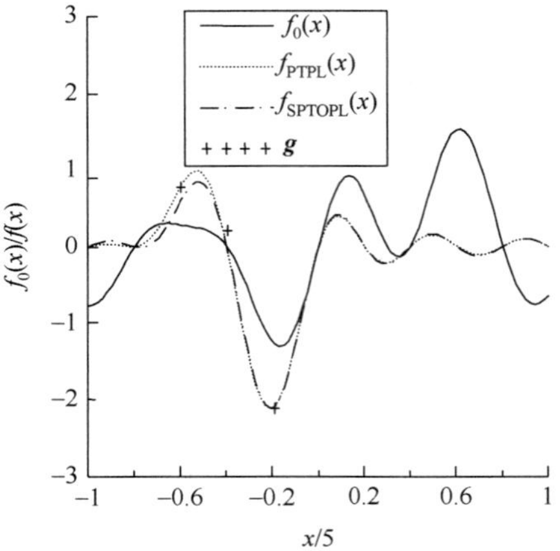

(b) 得到第三对训练数据时

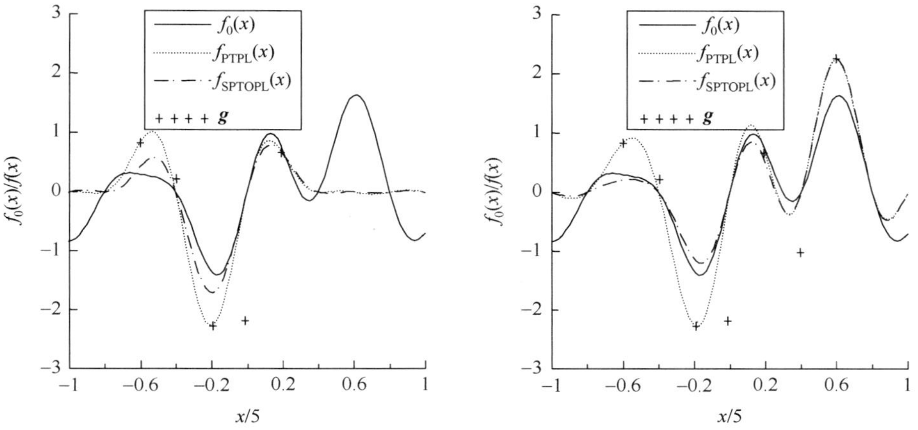


(c) 得到第五对训练数据时　　(d) 得到第七对训练数据时

图 3.5　基于观测值“＋”对未知信号[$f_0(x)$]的增量 PTPL 估计[$f_{PTPL}(x)$]、增量 SPTOPL 估计[$\mathcal{W}=\mathcal{A}^*$，$f_{SPTOPL}(x)$]

**表 3.2　未知信号 $f_0(x)$及其增量学习结果的坐标和方差**

| $f_0(x)$及其增量学习结果 | 坐标 | 方差 | 增量数据序号 |
|---|---|---|---|
| $f_0(x)$ | {−1.66，0.59，−2.67，1.43，3.25，−1.38} | | |
| $f_{PTPL}(x)$ | {0.00，1.65，0.00，0.00，0.00，0.00} | 0.58 | 1 |
| | {0.00，1.65，0.00，0.00，0.00，0.00} | 0.58 | 2 |
| | {0.00，1.65，−4.51，0.00，0.00，0.00} | 0.48 | 3 |
| | {0.00，1.65，−4.51，0.00，0.00，0.00} | 0.48 | 4 |
| | {0.00，1.65，−4.51，1.31，0.00，0.00} | 0.44 | 5 |
| | {0.00，1.65，−4.51，1.31，0.00，0.00} | 0.44 | 6 |
| | {0.00，1.65，−4.51，1.31，4.48，0.00} | 0.19 | 7 |
| $f_{SPTOPL}(x)$ | {0.00，1.65，0.00，0.00，0.00，0.00} | 0.58 | 1 |
| | {0.00，1.65，0.00，0.00，0.00，0.00} | 0.58 | 2 |
| | {0.00，1.24，−4.51，0.00，0.00，0.00} | 0.46 | 3 |

续表

| $f_0(x)$及其增量学习结果 | 坐标 | 方差 | 增量数据序号 |
|---|---|---|---|
| $f_{\mathrm{SPTOPL}}(x)$ | {0.00，1.24，−4.51，0.00，0.00，0.00} | 0.46 | 4 |
| | {0.00，0.82，−3.38，1.31，0.00，0.00} | 0.35 | 5 |
| | {0.00，0.82，−3.38，1.31，0.00，0.00} | 0.35 | 6 |
| | {0.00，0.41，−2.26，0.98，4.48，0.00} | 0.08 | 7 |

## 3.5 本章小结

本章以最优泛化为基本出发点讨论描述型投影学习。在介绍和总结描述型投影学习之多个具体基本准则和基本形式的基础上，进一步拓展得到对应扩展形式，并探讨了 S-L 投影学习和偏斜投影学习两种投影学习族，最后探讨了投影学习的增量形式，为在线学习、主动学习（训练样本选择）、模型稀疏化等应用奠定了基础。

# 第 4 章　鉴别型投影学习

## 4.1　引　　言

与描述型投影学习相对，鉴别型投影学习是通过学习来提取数据的鉴别性特征（discriminative features）或构建决策函数（decision function）即建立鉴别模型，称为鉴别子或鉴别器（discriminator）或分类器（classifier），完成模式识别这一机器学习的基本任务[252, 253, 425]，其基本性能要求是鉴别能力强。

本章延续第 3 章的基本思路，从基于逆问题求解的投影学习角度来讨论鉴别型学习，重点探讨核非线性鉴别子（kernel-based nonlinear discriminator，KND）、表示型核非线性鉴别子（kernel-based nonlinear representative discriminator，KNRD）、斜投影核鉴别子（kernel discriminator via oblique projection，KDOP）的投影学习。

## 4.2　核非线性鉴别子

本节主要讨论 KND 的学习准则、基本形式、投影学习机理与斜投影扩展、增量学习及稀疏化算法（即自适应训练）等内容。

### 4.2.1　KND 的学习准则与基本形式

在无噪标签的鉴别学习中，基于式（1.1）在对未知鉴别函数（鉴别子）$f_0(\boldsymbol{x})$的观测中无噪声影响，而按逆问题求解思路，求解结果$f(\boldsymbol{x})$的好坏取决于对式(3.10)中的学习子 $X$ 所实施的限制——学习准则。在 KND 中，对 $X$ 的基本限制是：按平均能量最小的要求抑制无关类别的影响，将感兴趣类别从这些类别中区分开来，这一准则有闭式解[426]。

1. KND 的学习准则：能量最小化

设共有 $C$ 个类别，对类别 $c(c = 1, 2, \cdots, C)$而言，需要找到学习子 $X^{(c)}$，以便能最大程度地将类别 $c$ 从其他类别中区分开来。为此，可以从抑制其他类别输出能量的角度，按如下准则寻找最优学习子[426]：

$$\mathcal{X}_{\text{opt,KND}}^{(c)} = \underset{\{\mathcal{X}^{(c)}\}}{\operatorname{argmin}} \left\{ \underset{i,\, i\neq c}{\text{mean}} \left\| \mathcal{X}^{(c)} \boldsymbol{g}^{(i)} \right\|^2 \right\}, \tag{4.1}$$

式中，$\boldsymbol{g}^{(i)}$为第 $i$ 类的标签向量$(i \neq c, i = 1, 2, \cdots, C)$。

这一准则建立了这样一个鉴别标准：除目标类别 $c$ 以外，压制其他类别的输出结果，使其平均能量最小。但对目标类别 $c$ 的输出没有任何限制，可以自由选择。

因此，该准则既可用于解决传统 LDA、KDA（kernel discriminant analysis，核鉴别分析）所解决的两类别鉴别问题[246, 427]，也可用于解决 PLS（partial least square，偏最小二乘）所解决的多类别鉴别问题[428, 429]。此外，目标类别 $c$ 在被视为离群类别、其他类别整体作为单个类别的条件下还可以用于解决背景建模、异常检测等应用场景中单类别鉴别问题，其中无须数据标签信息，也不存在不同类别训练样本量的平衡问题[430-432]。

若定义能量矩阵

$$\boldsymbol{Q}^{(i)} = \left( \boldsymbol{g}^{(i)} \otimes \overline{\boldsymbol{g}^{(i)}} \right), \tag{4.2}$$

则式（4.1）等价为

$$\mathcal{X}_{\text{opt,KND}}^{(c)} = \underset{\{\mathcal{X}^{(c)}\}}{\operatorname{argmin}} \left\{ \operatorname{tr}\left( \mathcal{X}^{(c)} \boldsymbol{Q}_{\text{KND}} \left[ \mathcal{X}^{(c)} \right]^* \right) \right\}, \tag{4.3}$$

其中

$$\boldsymbol{Q}_{\text{KND}} \equiv \frac{1}{C-1} \sum_{i=1,\neq c}^{C} \boldsymbol{Q}^{(i)} . \tag{4.4}$$

式（4.3）就是 KND 学习准则（学习子优化准则）。

### 2. KND 的基本形式与结构

利用文献[424]之定理 2.3.1，由式（4.3）和式（4.4）可以得到如下定理。

**定理 4.1（KND 的最优学习子）**对目标类别 $c$，KND 的最优学习子为

$$\mathcal{X}_{\text{opt,KND}}^{(c)} = \mathcal{G}(\boldsymbol{I}_N - \boldsymbol{Q}_{\text{KND}} \boldsymbol{Q}_{\text{KND}}^{+}), \tag{4.5}$$

式中，$\mathcal{G}$ 为从 $C^N$ 到 $H$ 的任意算子；$\boldsymbol{I}_N$ 为 $C^N$ 中的单位矩阵。

由于对目标类别 $c$ 的输出没有特定限制，所以原则上式（4.5）中 $\mathcal{G}$ 可以任意选择。为便于计算，这里特定地选取 $\mathcal{G} = [\mathcal{A}^{(c)}]^*$，即选取目标类别 $c$ 的样本按式(3.3)所确定观测算子之伴随算子。这时式（4.5）具体化为

$$\mathcal{X}_{\text{opt,KND}}^{(c)} = [\mathcal{A}^{(c)}]^* \left( \boldsymbol{I}_N - \boldsymbol{Q}_{\text{KND}} \boldsymbol{Q}_{\text{KND}}^{+} \right). \tag{4.6}$$

若 $H$ 是再生核为 $k$ 的 RKHS，式（3.3）中 $\psi_n(\boldsymbol{x})$取为 $\psi_n(\boldsymbol{x}) = k(\boldsymbol{x}, \boldsymbol{x}_n)$，则将式（4.6）中的学习子代入式（3.10），可得如下定理。

**定理 4.2（KND）**对目标类别 $c$，核非线性鉴别子为

$$f_{\mathrm{KND}}^{(c)}(\boldsymbol{x})=\sum_{j=1}^{N}a_{\mathrm{KND}}^{(c)}(j)k(\boldsymbol{x},\boldsymbol{x}_j^{(c)}),\tag{4.7}$$

其中，系数向量为

$$\boldsymbol{a}_{\mathrm{KND}}^{(c)}=\left[a_{\mathrm{KND}}^{(c)}(1),a_{\mathrm{KND}}^{(c)}(2),\cdots,a_{\mathrm{KND}}^{(c)}(N)\right]^{\mathrm{T}}=\left(\boldsymbol{I}_N-\boldsymbol{Q}_{\mathrm{KND}}\boldsymbol{Q}_{\mathrm{KND}}^{+}\right)\boldsymbol{g}^{(c)}.\tag{4.8}$$

显然，KND 与 KNR 具有相同的形式和结构（如图 3.1 所示，为单隐层网络）。

### 4.2.2　KND 的投影学习机理与斜投影扩展形式

本小节在讨论 KND 的投影学习机理之基础上，基于斜投影给出其扩展形式[433]。

1. KND 中的投影学习机理

由式（2.26）可知，式（4.8）中的 $\boldsymbol{Q}_{\mathrm{KND}}\boldsymbol{Q}_{\mathrm{KND}}^{+}$ 是 $\boldsymbol{Q}_{\mathrm{KND}}$ 之值域上的正投影[286]，即有 $\boldsymbol{Q}_{\mathrm{KND}}\boldsymbol{Q}_{\mathrm{KND}}^{+}=\mathcal{P}_{R(\boldsymbol{Q}_{\mathrm{KND}})}$。同时，由于 $\boldsymbol{Q}_{\mathrm{KND}}$ 是自伴算子，所以式（4.8）等价于

$$\boldsymbol{a}_{\mathrm{KND}}^{(c)}=\left[a_{\mathrm{KND}}^{(c)}(1),a_{\mathrm{KND}}^{(c)}(2),\cdots,a_{\mathrm{KND}}^{(c)}(N)\right]^{\mathrm{T}}=\left[\boldsymbol{I}_N-\mathcal{P}_{R(\boldsymbol{Q}_{\mathrm{KND}})}\right]\boldsymbol{g}^{(c)}=\mathcal{P}_{N(\boldsymbol{Q}_{\mathrm{KND}})}\boldsymbol{g}^{(c)},\tag{4.9}$$

即 KND 的系数向量是标签向量 $\boldsymbol{g}^{(c)}$在 $\boldsymbol{Q}_{\mathrm{KND}}$ 之零空间上的正投影。换言之，目标类别 $c$ 通过零化其他类别输出（使平均能量最小）的途径被鉴别出来，这一投影机理参见图 4.1[433]。

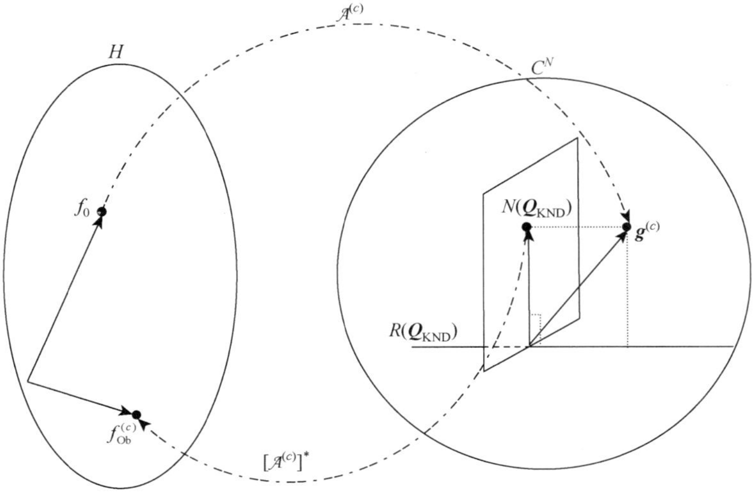


图 4.1　目标类别 $c$ 之 KND 的投影学习机理[433]

对比图 4.1 和图 3.2（b）可知，KND 在观测结果空间上保证最优鉴别能力，

不能像描述型投影学习表示子那样在原空间保证最优泛化能力，其泛化能力只能通过不断增加观测样本来提升。下一节介绍的 KND 之基于增量学习和稀疏化算法的自适应训练，正是逐渐提升其泛化能力的基本手段。

2. KND 的斜投影扩展形式[433]

显然，式（4.9）中 $\mathcal{P}_{N(\boldsymbol{Q}_{\text{KND}})}$ 与目标类别 $c$ 无关，若按式（4.2）定义其能量矩阵 $\boldsymbol{Q}^{(c)}$ 且有

$$\dim\left\{R(\boldsymbol{Q}^{(c)})\cup R(\boldsymbol{Q}_{\text{KND}})\right\}=\dim\left\{R(\boldsymbol{Q}^{(c)})\right\}+\dim\left\{R(\boldsymbol{Q}_{\text{KND}})\right\},\tag{4.10}$$

其中，dim 表示空间维数，则参考式（3.49）中的空间分解关系并由式（3.80）～式（3.83）可知，可以定义斜投影算子：

$$\mathcal{P}_{\text{ob}}=\mathcal{P}_{R(\boldsymbol{Q}^{(c)}),R(\boldsymbol{Q}_{\text{KND}})\cup\left\{R(\boldsymbol{Q}_{\text{KND}})\cup R(\boldsymbol{Q}^{(c)})\right\}^{\perp}},\tag{4.11}$$

以此计算 KND 的系数向量，将 KND 进行扩展，得到如下定理[433]。

**定理 4.3（eKND）** 对目标类别 $c$，扩展的核非线性鉴别子（extended KND，eKND）为

$$f_{\text{eKND}}^{(c)}(\boldsymbol{x})=\sum_{j=1}^{N}a_{\text{eKND}}^{(c)}(j)k(\boldsymbol{x},\boldsymbol{x}_j^{(c)}),\tag{4.12}$$

其中，系数向量为

$$\boldsymbol{a}_{\text{eKND}}^{(c)}=\left[a_{\text{eKND}}^{(c)}(1),a_{\text{eKND}}^{(c)}(2),\cdots,a_{\text{eKND}}^{(c)}(N)\right]^{\text{T}}=\mathcal{P}_{\text{Ob}}\boldsymbol{g}^{(c)}.\tag{4.13}$$

另外，容易证明式（4.11）中的斜投影算子可以按以下定理计算[433]。

**定理 4.4（eKND 的斜投影算子）** 目标类别 $c$ 的 eKND 斜投影算子可表示为

$$\mathcal{P}_{\text{ob}}=\mathcal{P}^{(c)}(\boldsymbol{I}_N-\mathcal{P})+\lambda\mathcal{P}^{(c)}\mathcal{P}\mathcal{P}^{(c)}(\boldsymbol{I}_N-\mathcal{P}),\tag{4.14}$$

其中

$$\mathcal{P}^{(c)}=\mathcal{P}_{R(\boldsymbol{Q}^{(c)})},\quad \mathcal{P}=\mathcal{P}_{R(\boldsymbol{Q}_{\text{KND}})},\quad \lambda=\left\|\boldsymbol{g}^{(c)}\right\|^2\Big/\left(\left\|\boldsymbol{g}^{(c)}\right\|^2-\left\langle\mathcal{P}\boldsymbol{g}^{(c)},\boldsymbol{g}^{(c)}\right\rangle\right),\tag{4.15}$$

事实上，由于 $\boldsymbol{Q}^{(c)}$ 是自伴算子且 $\mathcal{P}^{(c)}=\boldsymbol{Q}^{(c)}[\boldsymbol{Q}^{(c)}]^{+}$，所以由引理 3.7 和文献[416]可知，式（4.11）所对应的 eKND 之斜投影算子可表示为

$$\mathcal{P}_{\text{Ob}}=\sum_{\ell=0}^{\infty}(\mathcal{P}^{(c)}\mathcal{P})^{\ell}\mathcal{P}^{(c)}(\boldsymbol{I}_N-\mathcal{P}).\tag{4.16}$$

同时，由于 $\boldsymbol{Q}^{(c)}$ 按式（4.2）所得，所以有

$$\left[\boldsymbol{Q}^{(c)}\right]^{+}=\left[\boldsymbol{Q}^{(c)}\right]^{*}\left\{\boldsymbol{Q}^{(c)}\left[\boldsymbol{Q}^{(c)}\right]^{*}\right\}^{+}=\mathcal{P}^{(c)}\Big/\left\|\boldsymbol{g}^{(c)}\right\|^2.\tag{4.17}$$

另一方面，由文献[286]之式（3.11.9）可得

$$\left[\boldsymbol{Q}^{(c)}\right]^{+} = \boldsymbol{Q}^{(c)} \Big/ \left\|\boldsymbol{g}^{(c)}\right\|^{4}. \tag{4.18}$$

所以由式（4.2）、式（4.17）和式（4.18）可得

$$\mathcal{P}^{(c)} = \boldsymbol{Q}^{(c)} \Big/ \left\|\boldsymbol{g}^{(c)}\right\|^{2} = \left(\boldsymbol{g}^{(c)} \otimes \overline{\boldsymbol{g}^{(c)}}\right) \Big/ \left\|\boldsymbol{g}^{(c)}\right\|^{2}. \tag{4.19}$$

因此，由正投影算子的自伴性和诺伊曼-沙滕积（定义 2.15）及其性质可知

$$\mathcal{P}^{(c)}\mathcal{P} = \left(\boldsymbol{g}^{(c)} \otimes \overline{\mathcal{P}\boldsymbol{g}^{(c)}}\right) \Big/ \left\|\boldsymbol{g}^{(c)}\right\|^{2}, \tag{4.20}$$

进而有

$$\left(\mathcal{P}^{(c)}\mathcal{P}\right)^{\ell} = \left(\left\langle \mathcal{P}\boldsymbol{g}^{(c)}, \boldsymbol{g}^{(c)} \right\rangle \Big/ \left\|\boldsymbol{g}^{(c)}\right\|^{2}\right)^{\ell-1} \mathcal{P}^{(c)}\mathcal{P}, \quad \ell \geqslant 1. \tag{4.21}$$

最后，由式（4.15）、式（4.16）和式（4.21）可知，式（4.14）成立，故定理 4.4 成立。

定理 4.4 表明，eKND 的系数向量可分解为两个成分：

$$\boldsymbol{a}_{\mathrm{Ob}}^{(c)} = \boldsymbol{a}_{\mathrm{Ob},1}^{(c)} + \boldsymbol{a}_{\mathrm{Ob},2}^{(c)} = \boldsymbol{a}_{\mathrm{Ob},1}^{(c)} + \lambda\mathcal{P}^{(c)}\mathcal{P}\boldsymbol{a}_{\mathrm{Ob},1}^{(c)} = \left(I + \lambda\mathcal{P}^{(c)}\mathcal{P}\right)\boldsymbol{a}_{\mathrm{Ob},1}^{(c)}, \tag{4.22}$$

其中

$$\boldsymbol{a}_{\mathrm{Ob},1}^{(c)} = \mathcal{P}^{(c)}(\boldsymbol{I}_N - \mathcal{P})\boldsymbol{g}^{(c)} = \mathcal{P}^{(c)}\left(\boldsymbol{I}_N - \boldsymbol{Q}_{\mathrm{KND}}\boldsymbol{Q}_{\mathrm{KND}}^{+}\right)\boldsymbol{g}^{(c)} \tag{4.23}$$

正是作为 $\boldsymbol{g}^{(c)}$ 向 $N(\boldsymbol{Q}_{\mathrm{KND}})$ 正投影结果的 KND 系数向量再经 $\mathcal{P}^{(c)}$ 正投影到 $R(\boldsymbol{Q}^{(c)})$ 的结果，而 $\boldsymbol{a}_{\mathrm{Ob},2}^{(c)}$ 是 $\boldsymbol{a}_{\mathrm{Ob},1}^{(c)}$ 先后正投影到 $R(\boldsymbol{Q}_{\mathrm{KND}})$ 和 $R(\boldsymbol{Q}^{(c)})$ 上再伸展 $\lambda$ 倍的结果，如图 4.2 所示。

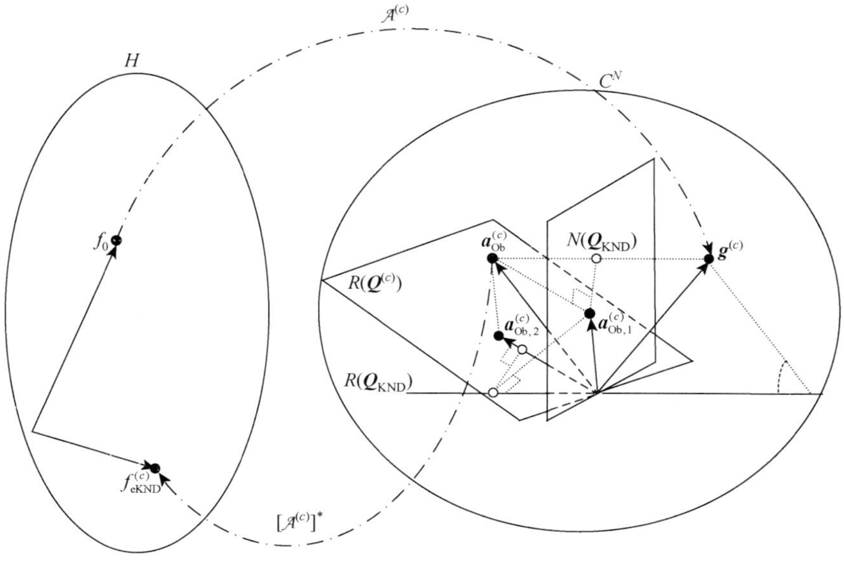


图 4.2　目标类别 $c$ 之 eKND 的投影学习机理

### 4.2.3　KND 的自适应训练

系统的自适应训练是根据环境和观测数据变化自动调整相应参数以提升其泛化能力的过程[418]，增量学习和稀疏化算法是实现自适应训练的重要手段。

增量学习在降低对应批量学习之内存消耗、提升模型稀疏性和表达能力时，往往以泛化能力为代价[339-342, 418-422]。与 3.4 节一样，这里讨论精确增量学习——学习结果与对应批量学习精确一致，不牺牲对应批量学习的泛化能力。同时，模型的稀疏化则是依据增量数据带来泛化能力为依据降低模型规模的过程。通过精确增量学习和稀疏化算法可以实现 KND 的自适应训练[360]，而 eKND 的自适应训练则可类似地完成。

1. KND 的增量学习算法

类似于 3.4 节，这里讨论 KND 的单样本增量学习算法。

对于目标类别 $c$，单样本增量学习旨在用已有系数向量 $\boldsymbol{a}_{\mathrm{KND},n}^{(c)}$ 和增量标签数据 $g_{n+1}^{(i)}(i=1,2,\cdots,C)$表示新的系数向量 $\boldsymbol{a}_{\mathrm{KND},n+1}^{(c)}$。类似于式（3.100）～式（3.111），先定义 $n$ 维向量：

$$\boldsymbol{q}_{\mathrm{KND},n+1}=\frac{1}{C-1}\sum_{i=1,i\neq c}^{C}\boldsymbol{g}_n^{(i)}\overline{g_{n+1}^{(i)}},\quad \boldsymbol{\tau}_{n+1}=\boldsymbol{Q}_{\mathrm{KND},n}^{+}\boldsymbol{q}_{\mathrm{KND},n+1},\tag{4.24}$$

以及标量

$$\sigma_{\mathrm{KND},n+1}=\frac{1}{C-1}\sum_{i=1,i\neq c}^{C}g_{n+1}^{(i)}\overline{g_{n+1}^{(i)}},\tag{4.25}$$

$$\alpha_{\mathrm{KND},n+1}=\sigma_{\mathrm{KND},n+1}-<\boldsymbol{\tau}_{n+1},\boldsymbol{q}_{\mathrm{KND},n+1}>,\tag{4.26}$$

$$\beta_{\mathrm{KND},n+1}=1+<\boldsymbol{\tau}_{n+1},\boldsymbol{\tau}_{n+1}>,\tag{4.27}$$

$$\gamma_{\mathrm{KND},n+1}=\left(g_{n+1}^{(c)}-<\boldsymbol{g}_n^{(c)},\boldsymbol{\tau}_{n+1}>\right)\Big/\beta_{\mathrm{KND},n+1}.\tag{4.28}$$

则有以下引理。

**引理 4.1**　对式（4.24）中的 $\boldsymbol{q}_{\mathrm{KND},n+1}$，有 $\boldsymbol{q}_{\mathrm{KND},n+1}\in R(\boldsymbol{Q}_{\mathrm{KND},n})$；对式（4.26）中的 $\alpha_{\mathrm{KND},n+1}$，有 $\alpha_{\mathrm{KND},n+1}\geqslant 0$。

事实上，由式（4.2）和式（4.4）可知，$\boldsymbol{Q}_{\mathrm{KND},n}$ 是非负定的，即 $\boldsymbol{Q}_{\mathrm{KND},n}\geqslant 0$。进一步由式（4.26）可得

$$\boldsymbol{Q}_{\mathrm{KND},n+1}=\begin{bmatrix}\boldsymbol{Q}_{\mathrm{KND},n} & \boldsymbol{q}_{\mathrm{KND},n+1}\\ \boldsymbol{q}_{\mathrm{KND},n+1}^{\mathrm{T}} & \sigma_{\mathrm{KND},n+1}\end{bmatrix}.\tag{4.29}$$

再由文献[434]之定理 1 和定理 2 可知，$\boldsymbol{Q}_{\mathrm{KND},n}\boldsymbol{Q}_{\mathrm{KND},n}^{+}\boldsymbol{q}_{\mathrm{KND},n+1}=\boldsymbol{q}_{\mathrm{KND},n+1}$、$\boldsymbol{\sigma}_{\mathrm{KND},n+1}\geqslant\ <\boldsymbol{Q}_{\mathrm{KND},n}^{+}\boldsymbol{q}_{\mathrm{KND},n+1},\boldsymbol{q}_{\mathrm{KND},n+1}>$，即有 $\boldsymbol{q}_{\mathrm{KND},n+1}\in R(\boldsymbol{Q}_{\mathrm{KND},n})$ 和 $\alpha_{\mathrm{KND},n+1}\geqslant 0$。

**引理 4.2（$\boldsymbol{Q}_{\mathrm{KND},n+1}$ 之广义逆的递归形式）** 对式（4.29）中的 $\boldsymbol{Q}_{\mathrm{KND},n+1}$ 有 $\boldsymbol{Q}_{\mathrm{KND},n+1}\geqslant 0$，且当 $\alpha_{\mathrm{KND},n+1}=0$ 时有

$$\boldsymbol{Q}_{\mathrm{KND},n+1}^{+}=\begin{bmatrix}\boldsymbol{T}_{n+1}\boldsymbol{Q}_{\mathrm{KND},n}^{+}\boldsymbol{T}_{n+1} & \boldsymbol{T}_{n+1}\boldsymbol{Q}_{\mathrm{KND},n}^{+}\boldsymbol{\tau}_{n+1}\big/\beta_{\mathrm{KND},n+1}\\ (\boldsymbol{T}_{n+1}\boldsymbol{Q}_{\mathrm{KND},n}^{+}\boldsymbol{\tau}_{n+1})^{\mathrm{T}}\big/\beta_{\mathrm{KND},n+1} & <\boldsymbol{Q}_{\mathrm{KND},n}^{+}\boldsymbol{\tau}_{n+1},\boldsymbol{\tau}_{n+1}>\big/\beta_{\mathrm{KND},n+1}^{2}\end{bmatrix},\tag{4.30}$$

其中

$$\boldsymbol{T}_{n+1}=\boldsymbol{I}_{n}-\left(\boldsymbol{\tau}_{n+1}\otimes\overline{\boldsymbol{\tau}_{n+1}}\right)\big/\beta_{\mathrm{KND},n+1}\,.\tag{4.31}$$

当 $\alpha_{\mathrm{KND},n+1}>0$ 时，有

$$\boldsymbol{Q}_{\mathrm{KND},n+1}^{+}=\begin{bmatrix}\boldsymbol{Q}_{\mathrm{KND},n}^{+}+\left(\boldsymbol{\tau}_{n+1}\otimes\overline{\boldsymbol{\tau}_{n+1}}\right)\big/\alpha_{\mathrm{KND},n+1} & -\boldsymbol{\tau}_{n+1}\big/\alpha_{\mathrm{KND},n+1}\\ -\boldsymbol{\tau}_{n+1}^{\mathrm{T}}\big/\alpha_{\mathrm{KND},n+1} & 1\big/\alpha_{\mathrm{KND},n+1}\end{bmatrix}.\tag{4.32}$$

事实上，由式（4.24）和引理 4.1 的证明可知，$\boldsymbol{Q}_{\mathrm{KND},n}\boldsymbol{Q}_{\mathrm{KND},n}^{+}\boldsymbol{q}_{\mathrm{KND},n+1}=\boldsymbol{q}_{\mathrm{KND},n+1}$，即 $\boldsymbol{Q}_{\mathrm{KND},n}\boldsymbol{\tau}_{n+1}=\boldsymbol{q}_{\mathrm{KND},n+1}$，且可知 $\alpha_{\mathrm{KND},n+1}\geqslant 0$。而由式（4.2）和式（4.4）可知，$\boldsymbol{Q}_{\mathrm{KND},n}\geqslant 0$。据此，由式（4.29）和文献[434]之定理 3 可知，引理 4.2 成立。

由引理 4.1 和引理 4.2，容易得到定理 4.5。

**定理 4.5（KND 的增量学习）**　KND 的系数有如下增量学习形式

$$\boldsymbol{a}_{\mathrm{KND},n+1}^{(c)}=\begin{cases}\begin{bmatrix}\boldsymbol{a}_{\mathrm{KND},n}^{(c)}-\gamma_{\mathrm{KND},n+1}\boldsymbol{\tau}_{n+1}\\ \gamma_{\mathrm{KND},n+1}\end{bmatrix}, & \alpha_{\mathrm{KND},n+1}=0,\\ \begin{bmatrix}\boldsymbol{a}_{\mathrm{KND},n}^{(c)}\\ 0\end{bmatrix}, & \alpha_{\mathrm{KND},n+1}>0.\end{cases}\tag{4.33}$$

事实上，由式（4.29）和引理 4.1、引理 4.2 可得

$$\boldsymbol{Q}_{\mathrm{KND},n+1}\boldsymbol{Q}_{\mathrm{KND},n+1}^{+}=\begin{cases}\begin{bmatrix}\boldsymbol{Q}_{\mathrm{KND},n}\boldsymbol{Q}_{\mathrm{KND},n}^{+}\boldsymbol{T}_{n+1} & \boldsymbol{\tau}_{n+1}\big/\beta_{\mathrm{KND},n+1}\\ \boldsymbol{\tau}_{n+1}^{\mathrm{T}}\big/\beta_{\mathrm{KND},n+1} & <\boldsymbol{\tau}_{n+1},\boldsymbol{\tau}_{n+1}>\big/\beta_{\mathrm{KND},n+1}\end{bmatrix}, & \alpha_{\mathrm{KND},n+1}=0,\\ \begin{bmatrix}\boldsymbol{Q}_{\mathrm{KND},n}\boldsymbol{Q}_{\mathrm{KND},n}^{+} & 0\\ 0 & 1\end{bmatrix}, & \alpha_{\mathrm{KND},n+1}>0,\end{cases}\tag{4.34}$$

其中，$\boldsymbol{T}_{n+1}$ 如式（4.31）。另外，由于

$$\boldsymbol{I}_{n+1}=\begin{bmatrix}\boldsymbol{I}_{n} & 0\\ 0 & 1\end{bmatrix},\quad \boldsymbol{g}_{n+1}^{(c)}=\begin{bmatrix}\boldsymbol{g}_{n}^{(c)}\\ g_{n+1}^{(c)}\end{bmatrix},\tag{4.35}$$

所以，将式（4.34）和式（4.35）代入式（4.8）即得式（4.33），因此定理 4.5 成立。

2. KND 的稀疏化

定理 4.5 表明，当第 $n+1$ 对增量数据到来以后，若 $\alpha_{\mathrm{KND},n+1}>0$，即 $\sigma_{\mathrm{KND},n+1}>\langle \boldsymbol{Q}_{\mathrm{KND},n}^{+}\boldsymbol{q}_{\mathrm{KND},n+1},\boldsymbol{q}_{\mathrm{KND},n+1}\rangle$，则模型系数不变，增量数据对提升鉴别能力和泛化能力没有带来任何收益，可以忽略不计（事实上，当收益不大时也可忽略），由此可实现模型的稀疏化。

对于目标类别 $c$，第 $n+1$ 对增量数据的影响可以用以下新息指标来刻画：

$$\delta_{n,n+1}^{(c)}=\left|f_{\mathrm{KND},n+1}^{(c)}(\boldsymbol{x}_{n+1}^{(c)})-f_{\mathrm{KND},n}^{(c)}(\boldsymbol{x}_{n+1}^{(c)})\right|. \tag{4.36}$$

显然，由定理 4.2 和定理 4.5 可知，该新息指标等价于：

$$\delta_{n,n+1}^{(c)}=\begin{cases}\left|\gamma_{\mathrm{KND},n+1}\right|\left|k(\boldsymbol{x}_{n+1}^{(c)},\boldsymbol{x}_{n+1}^{(c)})-\sum\limits_{j=1}^{n}\tau_{n+1}(j)k(\boldsymbol{x}_{n+1}^{(c)},\boldsymbol{x}_{j}^{(c)})\right|, & \alpha_{\mathrm{KND},n+1}=0,\\ 0, & \alpha_{\mathrm{KND},n+1}>0.\end{cases} \tag{4.37}$$

如果该指标值小于预设的门限，则第 $n+1$ 对增量数据可忽略不计。评估每一步增量学习的新息可实现 KND 的稀疏化。

## 4.3　表示型核非线性鉴别子

本节在 KND 的基础上引入最优泛化约束，将 KND 扩展成表示型核非线性鉴别子（kernel-based nonlinear representative discriminator，KNRD）。主要内容包括 KNRD 学习准则、基本形式和自适应训练。

### 4.3.1　KNRD 的投影学习准则和基本形式

对比式（4.1）和式（3.16）之第一项（最优泛化投影学习准则中的目标函数）可知，在 KND 中，目标类别 $c$ 以外其他类别的影响事实上被当作噪声处理。如果施加最优泛化投影约束（最优表示约束）并适当平衡鉴别能力和泛化能力，即可得到表示型鉴别子[396]。

事实上，记式（4.3）对应的 KND 学习子为

$$\mathcal{X}_{\mathrm{D}}^{(c)}=\underset{\{\mathcal{X}^{(c)}\}}{\operatorname{argmin}}\left\{\operatorname{tr}\left(\mathcal{X}^{(c)}\boldsymbol{Q}_{\mathrm{KND}}\left[\mathcal{X}^{(c)}\right]^{*}\right)\right\}, \tag{4.38}$$

同时，记式（3.16）之投影约束（第二项）对应学习子为

$$\mathcal{X}_{\mathrm{R}}^{(c)}=\underset{\{\mathcal{X}^{(c)}\}}{\operatorname{argmin}}\left\{\operatorname{tr}\left[\left(I-\mathcal{X}^{(c)}\mathcal{A}^{(c)}\right)\left(I-\mathcal{X}^{(c)}\mathcal{A}^{(c)}\right)^{*}\right]\right\}, \tag{4.39}$$

这就是最优表示（最优泛化）约束，其解对应于定理 3.1 中的 KNR。

若用参数$\lambda\,(0<\lambda<1)$控制式（4.38）之最优鉴别和式（4.39）之最优表示之间的平衡，则可以得到类似于式（3.23）的参数化投影学习准则，称为表示型最优鉴别（representative discrimination，RD）准则[396]：

$$\mathcal{X}_{\mathrm{RD}}^{(c)}=\underset{\{\mathcal{X}\}}{\operatorname{argmin}}\left\{\lambda\operatorname{tr}\left[\mathcal{X}^{(c)}\boldsymbol{Q}_{\mathrm{KND}}\left(\mathcal{X}^{(c)}\right)^{*}\right]+(1-\lambda)\operatorname{tr}\left[\left(I-\mathcal{X}^{(c)}\mathcal{A}^{(c)}\right)\left(I-\mathcal{X}^{(c)}\mathcal{A}^{(c)}\right)^{*}\right]\right\}. \tag{4.40}$$

所以，从本质上说，表示型最优鉴别准则是将目标类别以外的影响当成噪声处理时的参数化投影学习准则（参见图 3.2）。

对给定$\lambda\,(0<\lambda<1)$，式（4.40）等价于[396]

$$\mathcal{X}_{\mathrm{RD}}^{(c)}=\underset{\{\mathcal{X}\}}{\operatorname{argmin}}\left\{\rho\operatorname{tr}\left[\mathcal{X}^{(c)}\boldsymbol{Q}_{\mathrm{KND}}\left(\mathcal{X}^{(c)}\right)^{*}\right]+\operatorname{tr}\left[\left(I-\mathcal{X}^{(c)}\mathcal{A}^{(c)}\right)\left(I-\mathcal{X}^{(c)}\mathcal{A}^{(c)}\right)^{*}\right]\right\}, \tag{4.41}$$

其中，$\rho=\lambda/(1-\lambda)$。据此类似于定理 3.3，可以得到如下定理[396]。

**定理 4.6（KNRD）** 表示型最优鉴别学习子为

$$\mathcal{X}_{\mathrm{RD}}^{(c)}=\left(\mathcal{A}^{(c)}\right)^{*}\left(\boldsymbol{U}_{\mathrm{RD}}^{(c)}\right)^{+}+\mathcal{G}\left[\boldsymbol{I}_{N}-\boldsymbol{U}_{\mathrm{RD}}^{(c)}\left(\boldsymbol{U}_{\mathrm{RD}}^{(c)}\right)^{+}\right]. \tag{4.42}$$

对应地，对目标类别 $c$ 的核非线性表示型鉴别子（KNRD）为

$$f_{\mathrm{RD}}^{(c)}(\boldsymbol{x})=\sum_{j=1}^{N}a_{\mathrm{RD}}^{(c)}(j)k(\boldsymbol{x},\boldsymbol{x}_{j}^{(c)}), \tag{4.43}$$

其中，系数向量为

$$\boldsymbol{a}_{\mathrm{RD}}^{(c)}=\left[a_{\mathrm{RD}}^{(c)}(1),a_{\mathrm{RD}}^{(c)}(2),\cdots,a_{\mathrm{RD}}^{(c)}(N)\right]^{\mathrm{T}}=\left(\boldsymbol{U}_{\mathrm{RD}}^{(c)}\right)^{+}\boldsymbol{g}^{(c)}, \tag{4.44}$$

其中

$$\boldsymbol{U}_{\mathrm{RD}}^{(c)}=\boldsymbol{K}^{(c)}+\rho\boldsymbol{Q}_{\mathrm{KND}}, \tag{4.45}$$

其中，$\boldsymbol{K}^{(c)}$为目标类别训练样本对应的格拉姆矩阵。

显然，KNRD 与 KND 和 KNR 都有相同的结构（图 3.1，为单隐层网络）。

### 4.3.2　KNRD 的自适应训练

类似于 4.2.3 小节中对自适应 KND 的讨论，这里探讨基于增量学习和模型稀疏化的 KNRD 之自适应训练问题，简介如下[362]。

1. KNRD 的增量学习

对于目标类别 $c$，先定义 $n$ 维向量：

$$\boldsymbol{s}_{n+1}=\left[k(\boldsymbol{x}_{1}^{(c)},\boldsymbol{x}_{n+1}^{(c)}),k(\boldsymbol{x}_{2}^{(c)},\boldsymbol{x}_{n+1}^{(c)}),\cdots,k(\boldsymbol{x}_{n}^{(c)},\boldsymbol{x}_{n+1}^{(c)})\right]^{\mathrm{T}}, \tag{4.46}$$

$$\boldsymbol{t}_{\mathrm{RD},n+1}=\boldsymbol{s}_{n+1}+\rho\boldsymbol{q}_{\mathrm{KND},n+1}, \tag{4.47}$$

$$\widetilde{\boldsymbol{\tau}}_{n+1} = \left(\boldsymbol{U}_{\mathrm{RD},n}^{(c)}\right)^{+} \boldsymbol{t}_{\mathrm{RD},n+1}. \tag{4.48}$$

再定义标量：

$$\widetilde{\alpha}_{n+1} = k(\boldsymbol{x}_{n+1}^{(c)}, \boldsymbol{x}_{n+1}^{(c)}) + \rho\sigma_{\mathrm{KND},n+1} - < \widetilde{\boldsymbol{\tau}}_{n+1}, \boldsymbol{t}_{\mathrm{RD},n+1} >, \tag{4.49}$$

$$\widetilde{\beta}_{n+1} = 1 + < \widetilde{\boldsymbol{\tau}}_{n+1}, \widetilde{\boldsymbol{\tau}}_{n+1} >, \tag{4.50}$$

$$\widetilde{\gamma}_{n+1} = \left(g_{n+1}^{(c)} - < \boldsymbol{g}_n^{(c)}, \widetilde{\boldsymbol{\tau}}_{n+1} >\right) \Big/ \widetilde{\beta}_{n+1}. \tag{4.51}$$

则类似于引理 4.1 和引理 4.2，可得到如下结果：

（1）$\widetilde{\alpha}_{n+1} \geqslant 0$，$\boldsymbol{t}_{\mathrm{RD},n+1} \in R\left(\boldsymbol{U}_{\mathrm{RD},n}^{(c)}\right)$。

（2）当 $\widetilde{\alpha}_{n+1} = 0$ 时，有

$$\left(\boldsymbol{U}_{\mathrm{RD},n+1}^{(c)}\right)^{+} = \begin{bmatrix} \widetilde{\boldsymbol{T}}_{n+1}\left(\boldsymbol{U}_{\mathrm{RD},n}^{(c)}\right)^{+}\widetilde{\boldsymbol{T}}_{n+1} & \widetilde{\boldsymbol{T}}_{n+1}\left(\boldsymbol{U}_{\mathrm{RD},n}^{(c)}\right)^{+}\widetilde{\boldsymbol{\tau}}_{n+1} \Big/ \widetilde{\beta}_{n+1} \\ \left\{\widetilde{\boldsymbol{T}}_{n+1}\left(\boldsymbol{U}_{\mathrm{RD},n+1}^{(c)}\right)^{+}\widetilde{\boldsymbol{\tau}}_{n+1}\right\}^{\mathrm{T}} \Big/ \widetilde{\beta}_{n+1} & < \left(\boldsymbol{U}_{\mathrm{RD},n}^{(c)}\right)^{+}\widetilde{\boldsymbol{\tau}}_{n+1}, \widetilde{\boldsymbol{\tau}}_{n+1} > \Big/ \widetilde{\beta}_{n+1}^{2} \end{bmatrix}, \tag{4.52}$$

其中

$$\widetilde{\boldsymbol{T}}_{n+1} = \boldsymbol{I}_n - \left(\widetilde{\boldsymbol{\tau}}_{n+1} \otimes \overline{\widetilde{\boldsymbol{\tau}}_{n+1}}\right) \Big/ \widetilde{\beta}_{n+1}. \tag{4.53}$$

而当 $\widetilde{\alpha}_{n+1} > 0$ 时，有

$$\left(\boldsymbol{U}_{\mathrm{RD},n+1}^{(c)}\right)^{+} = \begin{bmatrix} \left(\boldsymbol{U}_{\mathrm{RD},n}^{(c)}\right)^{+} + \left(\widetilde{\boldsymbol{\tau}}_{n+1} \otimes \overline{\widetilde{\boldsymbol{\tau}}_{n+1}}\right) \Big/ \widetilde{\alpha}_{n+1} & -\widetilde{\boldsymbol{\tau}}_{n+1} \Big/ \widetilde{\alpha}_{n+1} \\ -\widetilde{\boldsymbol{\tau}}_{n+1}^{\mathrm{T}} \Big/ \widetilde{\alpha}_{n+1} & 1 \Big/ \widetilde{\alpha}_{n+1} \end{bmatrix}. \tag{4.54}$$

于是，类似于定理 4.5，得到如下定理。

**定理 4.7（KNRD 的增量学习）** KNRD 的系数有如下增量学习形式

$$\boldsymbol{a}_{\mathrm{RD},n+1}^{(c)} = \begin{cases} \begin{bmatrix} \boldsymbol{a}_{\mathrm{RD},n}^{(c)} - \eta_{n+1}\widetilde{\boldsymbol{\tau}}_{n+1} + \widetilde{\gamma}_{n+1}\left(\boldsymbol{U}_{\mathrm{RD},n}^{(c)}\right)^{+}\widetilde{\boldsymbol{\tau}}_{n+1} \\ \eta_{n+1} \end{bmatrix}, & \widetilde{\alpha}_{n+1} = 0, \\ \begin{bmatrix} \boldsymbol{a}_{\mathrm{RD},n}^{(c)} - \widetilde{\gamma}_{n+1}\widetilde{\beta}_{n+1}\widetilde{\boldsymbol{\tau}}_{n+1} \Big/ \widetilde{\alpha}_{n+1} \\ \widetilde{\gamma}_{n+1}\widetilde{\beta}_{n+1} \Big/ \widetilde{\alpha}_{n+1} \end{bmatrix}, & \widetilde{\alpha}_{n+1} > 0, \end{cases} \tag{4.55}$$

其中

$$\eta_{n+1} = \left[\widetilde{\gamma}_{n+1}\left\langle\left(\boldsymbol{U}_{\mathrm{RD},n}^{(c)}\right)^{+}\widetilde{\boldsymbol{\tau}}_{n+1}, \widetilde{\boldsymbol{\tau}}_{n+1}\right\rangle + \left\langle \widetilde{\boldsymbol{\tau}}_{n+1}, \boldsymbol{a}_n^{(c)}\right\rangle\right] \Big/ \widetilde{\beta}_{n+1}. \tag{4.56}$$

2. KNRD 的稀疏化

类似于式（4.37）的 KND 稀疏化指标，得到 KNRD 的稀疏化指标为

$$\widetilde{\delta}_{n,n+1}^{(c)}=\begin{cases}\left|\eta_{n+1}\right|\left|k(\boldsymbol{x}_{n+1}^{(c)},\boldsymbol{x}_{n+1}^{(c)})+\sum_{j=1}^{n}\widetilde{\nu}_{n+1}(j)k(\boldsymbol{x}_{n+1}^{(c)},\boldsymbol{x}_{j}^{(c)})\right|, & \widetilde{\alpha}_{n+1}=0,\\ \left|\dfrac{\widetilde{\gamma}_{n+1}\widetilde{\beta}_{n+1}}{\widetilde{\alpha}_{n+1}}\right|\left|k(\boldsymbol{x}_{n+1}^{(c)},\boldsymbol{x}_{n+1}^{(c)})+\sum_{j=1}^{n}\widetilde{\tau}_{n+1}(j)k(\boldsymbol{x}_{n+1}^{(c)},\boldsymbol{x}_{j}^{(c)})\right|, & \widetilde{\alpha}_{n+1}>0,\end{cases} \tag{4.57}$$

其中

$$\widetilde{\boldsymbol{\nu}}_{n+1}=\frac{\widetilde{\gamma}_{n+1}}{\eta_{n+1}}\left[\left(\boldsymbol{U}_{\mathrm{RD},n}^{(c)}\right)^{+}\widetilde{\boldsymbol{\tau}}_{n+1}-\widetilde{\boldsymbol{\tau}}_{n+1}\right]. \tag{4.58}$$

如果该指标值小于预设的门限，则第 $n+1$ 对增量数据可忽略不计。一旦在增量学习的每一步都对新息进行评估和处理，就可以实现 KNRD 的稀疏化。

## 4.4　斜投影核鉴别子

KND 和 eKND 都是在由观测样本决定的离散空间中设计的投影学习，其泛化能力不能从学习准则上得到保证，KNRD 则是最优泛化投影学习之参数化投影学习在将目标类别以外的类别视为噪声干扰时的特例，而且是基于正投影学习的鉴别子（分类器）。在大多数实际应用中，不同类别的特征所属子空间往往并不正交，因此有必要在原连续空间中引入基于斜投影学习的鉴别子，在确保泛化能力的同时便于解决一般场景下的鉴别问题。

本节以两类别鉴别问题为对象（多类别鉴别问题可以类似讨论），探讨斜投影核鉴别子（kernel-based discriminator via oblique projection，KDOP）的一般学习问题，主要内容包括 KDOP 学习的基本形式和增量形式两部分。

### 4.4.1　KDOP 的基本形式

无论是描述型投影学习还是鉴别型投影学习，都建立在基于式（1.1）之观测方程和基于式（3.10）之重构方程的逆问题求解之函数逼近框架上，其抽象如图 4.3 所示[211, 435]。本小节在这一框架内先给出 KDOP 的基本形式，再给出相关证明。

1. KDOP：斜投影核鉴别子

为简单起见，考虑两类别鉴别问题，用上标（1）和（2）分别表示两类别的训练数据集，即 $G^{(1)}=\{\boldsymbol{x}_i^{(1)},g_i^{(1)}\}_{i=1}^{N_1}$ 和 $G^{(2)}=\{\boldsymbol{x}_j^{(2)},g_j^{(2)}\}_{j=1}^{N_2}$。其中，$N_1$ 和 $N_2$ 分别为二者的训练样本数；$\boldsymbol{x}_i^{(1)}$、$\boldsymbol{x}_j^{(2)}$均为 $M$ 维向量，即 $\boldsymbol{x}_i^{(1)}\in C^M$、$\boldsymbol{x}_j^{(2)}\in C^M$。同样，假设鉴别子 $f_0(\boldsymbol{x})$所属原空间 $H$ 是具有再生核 $k$ 的 RKHS，则有

$$S_c = \text{span}\{k(\boldsymbol{x}, \boldsymbol{x}_1^{(c)}), k(\boldsymbol{x}, \boldsymbol{x}_2^{(c)}), \cdots, k(\boldsymbol{x}, \boldsymbol{x}_{N_c}^{(c)})\} \subset H,\ c = 1, 2, \tag{4.59}$$

也就是说，$S_1$ 和 $S_2$ 分别是由样本点集 $\{\boldsymbol{x}_i^{(1)}\}_{i=1}^{N_1}$ 和 $\{\boldsymbol{x}_j^{(2)}\}_{j=1}^{N_2}$ 决定的 $H$ 之子空间，并设 $f_0(x) \in S = S_1 \dotplus S_2 \subset H$ 。则用 $f_0(\boldsymbol{x})$的沿 $S_1$ 向 $S_2$ 之斜投影

$$f(\boldsymbol{x}) = \mathcal{P}_{S_2, S_1} f_0(\boldsymbol{x}) \tag{4.60}$$

将类别（2）从类别（1）中鉴别开来，如图 4.4[436-438]所示。

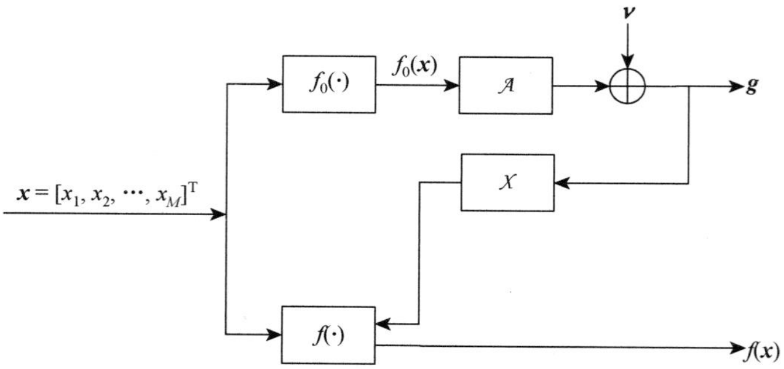


图 4.3　投影学习的函数逼近框架：逆问题求解[211, 435]

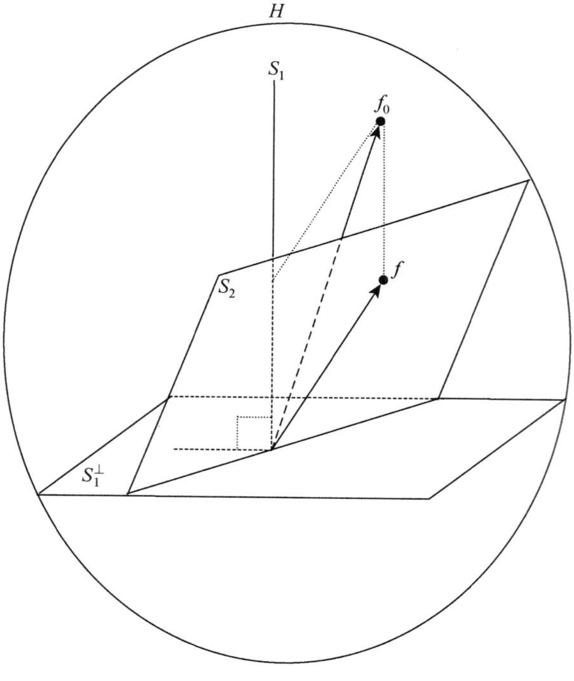


图 4.4　$f_0$ 沿 $S_1$ 向 $S_2$ 的斜投影[436-438]

进一步，由式（3.3）可得

$$\mathcal{A}_c=\sum_{n=1}^{N_c}\left(\boldsymbol{e}_n\otimes\overline{k(\boldsymbol{x},\boldsymbol{x}_n^{(c)})}\right),\ c=1,2\ . \tag{4.61}$$

考虑无噪标签问题（对于含噪标签问题，可以参照 3.3.3 小节中偏斜投影学习加以讨论），则由式（1.1）可得

$$\boldsymbol{g}^{(c)}=\mathcal{A}_c f_0(\boldsymbol{x})=\sum_{n=1}^{N_c}< f_0(\boldsymbol{x}),k(\boldsymbol{x},\boldsymbol{x}_n^{(c)})>\boldsymbol{e}_n,\ c=1,2\ . \tag{4.62}$$

另外，设由样本点集 $\{\boldsymbol{x}_i^{(1)}\}_{i=1}^{N_1}$、$\{\boldsymbol{x}_j^{(2)}\}_{j=1}^{N_2}$ 和再生核 $k$ 参照式（3.13）得到格拉姆矩阵分别为 $\boldsymbol{K}_1$ 和 $\boldsymbol{K}_2$，并记由向量

$$\boldsymbol{c}_j=\left[k(\boldsymbol{x}_j^{(2)},\boldsymbol{x}_1^{(1)}),\quad k(\boldsymbol{x}_j^{(2)},\boldsymbol{x}_2^{(1)}),\quad\cdots,\quad k(\boldsymbol{x}_j^{(2)},\boldsymbol{x}_{N_1}^{(1)})\right]^{\mathrm{T}},\ j=1,2,\cdots,N_2 \tag{4.63}$$

构成的矩阵为

$$\boldsymbol{C}=\left[\boldsymbol{c}_1,\quad \boldsymbol{c}_2,\quad \cdots,\quad \boldsymbol{c}_{N_2}\right]. \tag{4.64}$$

则有如下定理[436]。

**定理 4.8（斜投影核鉴别子，KDOP）**　将类别（2）从类别（1）中鉴别开来的斜投影核鉴别子为

$$f_{\mathrm{KDOP}}(\boldsymbol{x})=\mathcal{P}_{S_2,S_1}f_0(\boldsymbol{x})=\sum_{j=1}^{N_2}[b_{\mathrm{KDOP}}(j)-a_{\mathrm{KDOP}}(j)]k(\boldsymbol{x},\boldsymbol{x}_j^{(2)}), \tag{4.65}$$

式中的系数向量为

$$\boldsymbol{b}_{\mathrm{KDOP}}=[b_{\mathrm{KDOP}}(1),b_{\mathrm{KDOP}}(2),\cdots,b_{\mathrm{KDOP}}(N_2)]^{\mathrm{T}}=\boldsymbol{G}\boldsymbol{g}^{(2)}, \tag{4.66}$$

$$\boldsymbol{a}_{\mathrm{KDOP}}=[a_{\mathrm{KDOP}}(1),a_{\mathrm{KDOP}}(2),\cdots,a_{\mathrm{KDOP}}(N_2)]^{\mathrm{T}}=\boldsymbol{G}\boldsymbol{C}^{\mathrm{T}}\boldsymbol{K}_1^{+}\boldsymbol{g}^{(1)}, \tag{4.67}$$

其中

$$\boldsymbol{G}=\boldsymbol{K}_{1,\perp}^{+},\quad \boldsymbol{K}_{1,\perp}=\boldsymbol{K}_2-\boldsymbol{C}^{\mathrm{T}}\boldsymbol{K}_1^{+}\boldsymbol{C}\ . \tag{4.68}$$

2. 定理 4.8 的证明

首先，根据式（2.14）、式（2.16）和式（2.18），由式（4.59）可知 $H$ 中 $S_1$ 上的正投影算子为

$$\mathcal{P}_{S_1}=\sum_{n=1}^{N_1}\left(k(\boldsymbol{x},\boldsymbol{x}_n^{(1)})\otimes\overline{\widetilde{k(\boldsymbol{x},\boldsymbol{x}_n^{(1)})}}\right), \tag{4.69}$$

其中

$$\widetilde{k(\boldsymbol{x},\boldsymbol{x}_n^{(1)})}=\sum_{i=1}^{N_1}\alpha_{ni}k(\boldsymbol{x},\boldsymbol{x}_i^{(1)}),\ \alpha_{ni}=\left[\boldsymbol{K}_1^{+}\right]_{ni},n=1,2,\cdots,N_1. \tag{4.70}$$

其次，若令

$$\varphi_j(\boldsymbol{x}) = k(\boldsymbol{x}, \boldsymbol{x}_j^{(2)}) - \mathcal{P}_{S_1} k(\boldsymbol{x}, \boldsymbol{x}_j^{(2)}),\ j = 1, 2, \cdots, N_2, \tag{4.71}$$

则有

$$S_{1,\perp} = S_1^{\perp} = \operatorname{span}\{\varphi_1(\boldsymbol{x}), \varphi_2(\boldsymbol{x}), \cdots, \varphi_{N_2}(\boldsymbol{x})\}. \tag{4.72}$$

由此可得对应格拉姆矩阵及其广义逆矩阵

$$\boldsymbol{K}_{1,\perp} = \left[<\varphi_i(\boldsymbol{x}), \varphi_j(\boldsymbol{x})>\right], \boldsymbol{G} = \boldsymbol{K}_{1,\perp}^{+} = \left[\beta_{ij}\right], \tag{4.73}$$

以及[363]

$$\mathcal{P}_{S_2,S_1} = \sum_{j=1}^{N_2}\left(k(\boldsymbol{x}, \boldsymbol{x}_j^{(2)}) \otimes \overline{\widetilde{\varphi_j(\boldsymbol{x})}}\right), \tag{4.74}$$

其中

$$\widetilde{\varphi_j(\boldsymbol{x})} = \sum_{i=1}^{N_2}\beta_{ji}\varphi_i(\boldsymbol{x}). \tag{4.75}$$

将式（4.71）和式（4.75）代入式（4.74），并考虑到式（4.73）中 $G$ 的自伴性，可得

$$\mathcal{P}_{S_2,S_1} = \sum_{i=1}^{N_2}\sum_{j=1}^{N_2}\left(\beta_{ij} k(\boldsymbol{x}, \boldsymbol{x}_i^{(2)}) \otimes \left[\overline{\widetilde{k(\boldsymbol{x}, \boldsymbol{x}_j^{(2)})}} - \overline{\widetilde{\mathcal{P}_{S_1} k(\boldsymbol{x}, \boldsymbol{x}_j^{(2)})}}\right]\right). \tag{4.76}$$

另外，由式（4.69）和式（4.70）可得

$$\mathcal{P}_{S_1} k(\boldsymbol{x}, \boldsymbol{x}_j^{(2)}) = \sum_{m=1}^{N_1}\sum_{n=1}^{N_1}\alpha_{mn} c_j(n) k(\boldsymbol{x}, \boldsymbol{x}_m^{(1)}), \tag{4.77}$$

其中，$c_j(i)$为式（4.63）中向量 $\boldsymbol{c}_j$ 的第 $i$ 个元素。将式（4.77）代入式（4.76）可得

$$\mathcal{P}_{S_2,S_1} = \sum_{i=1}^{N_2}\sum_{j=1}^{N_2}\left(\beta_{ij} k(\boldsymbol{x}, \boldsymbol{x}_i^{(2)}) \otimes \left[\overline{\widetilde{k(\boldsymbol{x}, \boldsymbol{x}_j^{(2)})}} - \overline{\sum_{n=1}^{N_1}\sum_{m=1}^{N_1}\alpha_{nm} c_j(m) k(\boldsymbol{x}, \boldsymbol{x}_n^{(1)})}\right]\right), \tag{4.78}$$

由此得到

$$\begin{aligned} f_{\mathrm{KDOP}}(\boldsymbol{x}) &= \mathcal{P}_{S_2,S_1} f_0(\boldsymbol{x}) \\ &= \sum_{i=1}^{N_2}\left[\sum_{j=1}^{N_2}\left(\beta_{ij} < f_0(\boldsymbol{x}), k(\boldsymbol{x}, \boldsymbol{x}_j^{(2)}) - \sum_{n=1}^{N_1}\sum_{m=1}^{N_1}\alpha_{nm} c_j(m) k(\boldsymbol{x}, \boldsymbol{x}_n^{(1)}) >\right)\right] k(\boldsymbol{x}, \boldsymbol{x}_i^{(2)}), \end{aligned} \tag{4.79}$$

即式（4.65）成立，其中

$$\begin{aligned} \boldsymbol{b}_{\mathrm{KDOP}} &= [b_{\mathrm{KDOP}}(1), b_{\mathrm{KDOP}}(2), \cdots, b_{\mathrm{KDOP}}(N_2)]^{\mathrm{T}} = \sum_{j=1}^{N_2}\beta_{ij} < f_0(\boldsymbol{x}), k(\boldsymbol{x}, \boldsymbol{x}_j^{(2)})> \\ &= \sum_{j=1}^{N_2}\beta_{ij} g_j^{(2)} = \boldsymbol{G}\boldsymbol{g}^{(2)}, \end{aligned} \tag{4.80}$$

$$\boldsymbol{a}_{\mathrm{KDOP}}=[a_{\mathrm{KDOP}}(1),a_{\mathrm{KDOP}}(2),\cdots,a_{\mathrm{KDOP}}(N_2)]^{\mathrm{T}}=\sum_{j=1}^{N_2}\left(\sum_{m=1}^{N_1}\sum_{n=1}^{N_1}\beta_{ij}\alpha_{mn}c_j(n)g_m^{(1)}\right) \\ =\boldsymbol{GC}^{\mathrm{T}}\boldsymbol{K}_1^{+}\boldsymbol{g}^{(1)}. \tag{4.81}$$

于是，定理 4.8 得证。

由定理可见，KDOP 的投影机理如图 4.5 所示，其等价的投影合成机理如图 4.6[436-438]所示。

值得注意的是，当 $S_2$ 与 $S_1$ 正交时 KDOP 成为 KNR（这时 $S_2=S_1^{\perp}$，$\boldsymbol{C}=\boldsymbol{0}$，$\boldsymbol{K}_{1,\perp}=\boldsymbol{K}_2$）。

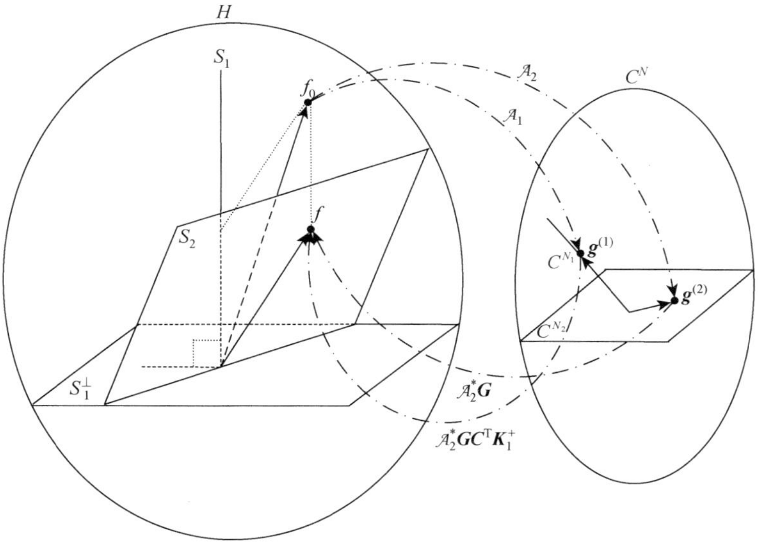


图 4.5　KDOP 的投影机理[436-438]

### 4.4.2　KDOP 的增量形式

由于 KDOP 涉及多类别训练数据，而不同类别的已有样本数不同，任何类别出现增量数据后都需要动态调整 KDOP 的参数。因此，在讨论 KDOP 的增量形式时需要对不同类别的增量数据加以区分，这里用上标、下标表示可变训练集元素和处理结果，其具体意义均可由前后关联性一目了然。例如，对两类别而言，设（1）类和（2）类已有样本数分别为可变数 $N_1=m$ 和 $N_2=n$，用 $\boldsymbol{g}_m^{(1)}$ 表示第（1）类的 $m$ 个标签构成的标签向量，用 $\boldsymbol{x}_{m+1}^{(1)}$ 表示第（1）类的第 $m+1$ 个训练样本，而用 $\boldsymbol{a}^{(m+1,n)}$表示第（1）、（2）类分别有 $m+1$ 个和 $n$ 个训练样本时得到的参数向量，而第（1）类和第（2）类已有样本数分别为 $m$ 和 $n$ 时类别（2）的 KDOP 为

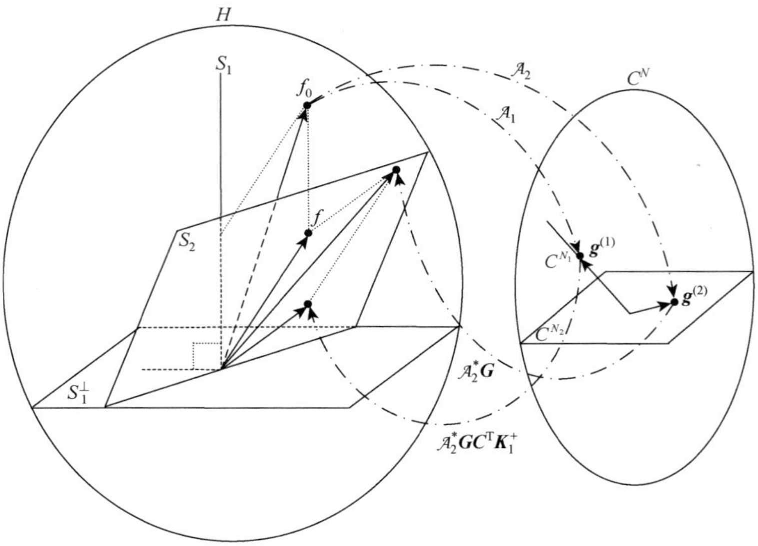


图 4.6　KDOP 的投影合成机理[436-438]

$$f_{\mathrm{KDOP}}^{(m,n)}(\boldsymbol{x})=\sum_{i=1}^{n}\left[b_{\mathrm{KDOP}}^{(m,n)}(i)-a_{\mathrm{KDOP}}^{(m,n)}(i)\right]k(\boldsymbol{x},\boldsymbol{x}_i^{(2)}),\tag{4.82}$$

其中

$$\boldsymbol{b}_{\mathrm{KDOP}}^{(m,n)}=\left[b_{\mathrm{KDOP}}^{(m,n)}(1),\ b_{\mathrm{KDOP}}^{(m,n)}(2),\ \cdots,\ b_{\mathrm{KDOP}}^{(m,n)}(n)\right]^{\mathrm{T}}=\boldsymbol{G}^{(m,n)}\boldsymbol{g}_n^{(2)},\tag{4.83}$$

$$\boldsymbol{a}_{\mathrm{KDOP}}^{(m,n)}=\left[a_{\mathrm{KDOP}}^{(m,n)}(1),\ a_{\mathrm{KDOP}}^{(m,n)}(2),\ \cdots,\ a_{\mathrm{KDOP}}^{(m,n)}(n)\right]^{\mathrm{T}}=\boldsymbol{G}^{(m,n)}\boldsymbol{C}_{(m,n)}^{\mathrm{T}}\boldsymbol{K}_{1,m}^{+}\boldsymbol{g}_m^{(1)}.\tag{4.84}$$

本小节以此约定讨论 KDOP 的增量学习和稀疏化问题，并对（1）、（2）类增量数据分别加以描述，即分别讨论 $G^{(1)}=\{\boldsymbol{x}_i^{(1)},g_i^{(1)}\}_{i=1}^{m+1}$、$G^{(2)}=\{\boldsymbol{x}_j^{(2)},g_j^{(2)}\}_{j=1}^{n+1}$ 对模型的影响，具体体现为 $\{\boldsymbol{x}_{m+1}^{(1)},g_{m+1}^{(1)}\}$ 和 $\{\boldsymbol{x}_{n+1}^{(2)},g_{n+1}^{(2)}\}$ 对 KDOP 模型参数的修正情况。

### 1. KDOP 的增量学习与稀疏化：对第（1）类样本

对第（1）类样本，考虑在训练集 $G^{(1)}=\{\boldsymbol{x}_i^{(1)},g_i^{(1)}\}_{i=1}^{m}$ 的基础上增加 $\{\boldsymbol{x}_{m+1}^{(1)},g_{m+1}^{(1)}\}$ 时的影响，即 $f_{\mathrm{KDOP}}^{(m+1,n)}(\boldsymbol{x})$ 与 $f_{\mathrm{KDOP}}^{(m,n)}(\boldsymbol{x})$ 的关系，体现为系数向量是否修正和如何修正。增量学习旨在将 $f_{\mathrm{KDOP}}^{(m+1,n)}(\boldsymbol{x})$ 表示为用 $\{\boldsymbol{x}_{m+1}^{(1)},g_{m+1}^{(1)}\}$ 对 $f_{\mathrm{KDOP}}^{(m,n)}(\boldsymbol{x})$ 加以修正的形式。

首先，由式（4.62）容易得到标签向量 $\boldsymbol{g}_{m+1}^{(1)}$ 的增量形式：

$$\boldsymbol{g}_{m+1}^{(1)}=\mathcal{A}_{1,m+1}f_0(\boldsymbol{x})=\sum_{i=1}^{m+1}<f_0(\boldsymbol{x}),k(\boldsymbol{x},\boldsymbol{x}_i^{(1)})>\boldsymbol{e}_i=\left[\{\boldsymbol{g}_m^{(1)}\}^{\mathrm{T}},g_{m+1}^{(1)}\right]^{\mathrm{T}}.\tag{4.85}$$

而由式（4.63）可得

$$\boldsymbol{c}_{m+1,j}=\left[k(\boldsymbol{x}_j^{(2)},\boldsymbol{x}_1^{(1)}),\quad k(\boldsymbol{x}_j^{(2)},\boldsymbol{x}_2^{(1)}),\quad\cdots,\quad k(\boldsymbol{x}_j^{(2)},\boldsymbol{x}_{m+1}^{(1)})\right]^{\mathrm{T}},\ j=1,2,\cdots,n,\tag{4.86}$$

由此可得 $\boldsymbol{C}_{m+1,n}$ 的增量形式：

$$\boldsymbol{C}_{m+1,n}=\left[\boldsymbol{c}_{m+1,1},\boldsymbol{c}_{m+1,2},\cdots,\boldsymbol{c}_{m+1,n}\right]=\begin{bmatrix}\boldsymbol{C}_{m,n}\\ \boldsymbol{d}_{m+1,n}^{\mathrm{T}}\end{bmatrix}=\left[\boldsymbol{C}_{m,n}^{\mathrm{T}},\boldsymbol{d}_{m+1,n}\right]^{\mathrm{T}}, \tag{4.87}$$

其中

$$\boldsymbol{d}_{m+1,n}\equiv\mathcal{A}_{2,n}k(\boldsymbol{x},\boldsymbol{x}_{m+1}^{(1)})=\left[k(\boldsymbol{x}_{m+1}^{(1)},\boldsymbol{x}_1^{(2)}),k(\boldsymbol{x}_{m+1}^{(1)},\boldsymbol{x}_2^{(2)}),\cdots,k(\boldsymbol{x}_{m+1}^{(1)},\boldsymbol{x}_n^{(2)})\right]^{\mathrm{T}}, \tag{4.88}$$

其次，参照式（4.52）和式（4.54）可得 $\boldsymbol{K}_{1,m+1}^{+}$ 的增量形式[361]：

$$\boldsymbol{K}_{1,m+1}^{+}=\begin{cases}\begin{bmatrix}\boldsymbol{K}_{1,m}^{+}, & \boldsymbol{0}_{m\times 1}\\ \boldsymbol{0}_{1\times m}, & 0\end{bmatrix}, & \lambda_{1,m+1}=0,\\ \begin{bmatrix}\boldsymbol{K}_{1,m}^{+}+\dfrac{\boldsymbol{\tau}_{1,m+1}\boldsymbol{\tau}_{1,m+1}^{\mathrm{T}}}{\lambda_{m+1}}, & -\dfrac{\boldsymbol{\tau}_{1,m+1}}{\lambda_{1,m+1}}\\ -\dfrac{\boldsymbol{\tau}_{1,m+1}^{\mathrm{T}}}{\lambda_{1,m+1}}, & \dfrac{1}{\lambda_{1,m+1}}\end{bmatrix}, & \lambda_{1,m+1}>0,\end{cases} \tag{4.89}$$

其中

$$\lambda_{1,m+1}=k(\boldsymbol{x}_{m+1}^{(1)},\boldsymbol{x}_{m+1}^{(1)}),\quad \boldsymbol{\tau}_{1,m+1}=\boldsymbol{K}_{1,m}^{+}\xi_{1,m+1}, \tag{4.90}$$

$$\xi_{1,m+1}\equiv\mathcal{A}_{1,m}k(\boldsymbol{x},\boldsymbol{x}_{m+1}^{(1)})=\left[k(\boldsymbol{x}_{m+1}^{(1)},\boldsymbol{x}_1^{(1)}),k(\boldsymbol{x}_{m+1}^{(1)},\boldsymbol{x}_2^{(1)}),\cdots,k(\boldsymbol{x}_{m+1}^{(1)},\boldsymbol{x}_m^{(1)})\right]^{\mathrm{T}}. \tag{4.91}$$

进而由式（4.87）和式（4.89）可得

$$\boldsymbol{C}_{m+1,n}^{\mathrm{T}}\boldsymbol{K}_{1,m+1}^{+}\boldsymbol{C}_{m+1,n}=\begin{cases}\boldsymbol{C}_{m,n}^{\mathrm{T}}\boldsymbol{K}_{1,m}^{+}\boldsymbol{C}_{m+1,n}, & \lambda_{1,m+1}=0,\\ \boldsymbol{C}_{m,n}^{\mathrm{T}}\boldsymbol{K}_{1,m+1}^{+}\boldsymbol{C}_{m+1,n}+\boldsymbol{\mu}_{1,m+1}\boldsymbol{\mu}_{1,m+1}^{\mathrm{T}}, & \lambda_{1,m+1}>0,\end{cases} \tag{4.92}$$

其中

$$\boldsymbol{\mu}_{1,m+1}=\frac{\boldsymbol{d}_{m+1}^{(n)}-\boldsymbol{C}_{m,n}^{\mathrm{T}}\boldsymbol{\tau}_{1,m+1}}{\lambda_{1,m+1}^{1/2}}. \tag{4.93}$$

结合式（4.68）和式（4.92）可得

$$\boldsymbol{K}_{1,\perp,m+1,n}=\boldsymbol{K}_{2,n}-\boldsymbol{C}_{m+1,n}^{\mathrm{T}}\boldsymbol{K}_{1,m+1}^{+}\boldsymbol{C}_{m+1,n}=\begin{cases}\boldsymbol{K}_{1,\perp,m+1,n}, & \lambda_{1,m+1}=0,\\ \boldsymbol{K}_{1,\perp,m+1,n}-\boldsymbol{\mu}_{1,m+1}\boldsymbol{\mu}_{1,m+1}^{\mathrm{T}}, & \lambda_{1,m+1}>0.\end{cases} \tag{4.94}$$

而由式（4.87）和式（4.88）可知

$$\boldsymbol{d}_{m+1,n}\in R(\mathcal{A}_{2,n})=R(\mathcal{A}_{2,n}\mathcal{A}_{2,n}^{*})\equiv R(\boldsymbol{K}_{2,n}), \tag{4.95}$$

$$\boldsymbol{C}_{m,n}^{\mathrm{T}}\in R(\mathcal{A}_{2,n})\equiv R(\boldsymbol{K}_{2,n})\Rightarrow \boldsymbol{C}_{m,n}^{\mathrm{T}}\boldsymbol{\tau}_{1,m+1}\in R(\mathcal{A}_{2,n})\equiv R(\boldsymbol{K}_{2,n}). \tag{4.96}$$

进而，由式（4.94）可知，对式（4.93）中定义的 $\boldsymbol{\mu}_{1,m+1}$，有

$$\boldsymbol{\mu}_{1,m+1}\in R(\mathcal{A}_{2,n})\equiv R(\boldsymbol{K}_{2,n})\Rightarrow \boldsymbol{\mu}_{1,m+1}\in R(\boldsymbol{K}_{1,\perp,m+1,n}). \tag{4.97}$$

由式（4.94）、式（4.97）以及文献[286]之定理 4.6 可得 $\boldsymbol{G}^{(m+1,n)}$的增量形式：

$$\boldsymbol{G}^{(m+1,n)} = \boldsymbol{K}_{1,\perp,m+1,n}^{+} = \begin{cases} \boldsymbol{G}^{(m,n)}, & \lambda_{1,m+1} = 0, \\ \left(\boldsymbol{I}_n + \boldsymbol{\theta}_{m+1}\boldsymbol{\zeta}_{1,m+1}^{\mathrm{T}}\right)\boldsymbol{G}^{(m,n)}, & \lambda_{1,m+1} > 0, \end{cases} \tag{4.98}$$

其中

$$\boldsymbol{\theta}_{m+1} = \boldsymbol{G}^{(m,n)}\boldsymbol{\mu}_{1,m+1}, \quad \boldsymbol{\zeta}_{1,m+1} = \left[1 - \boldsymbol{\mu}_{1,m+1}^{\mathrm{T}}\boldsymbol{\varepsilon}_{m+1}\right]^{-1}\boldsymbol{\mu}_{1,m+1} . \tag{4.99}$$

最后，得到系数向量 $\boldsymbol{b}_{\mathrm{KDOP}}^{(m+1,n)}$ 的增量形式：

$$\boldsymbol{b}_{\mathrm{KDOP}}^{(m+1,n)} = \boldsymbol{G}^{(m+1,n)}\boldsymbol{g}_n^{(2)} = \begin{cases} \boldsymbol{b}_{\mathrm{KDOP}}^{(m,n)}, & \lambda_{1,m+1} = 0, \\ \boldsymbol{b}_{\mathrm{KDOP}}^{(m,n)} + \delta_{1,b}\boldsymbol{\theta}_{m+1}, & \lambda_{1,m+1} > 0, \end{cases} \tag{4.100}$$

其中

$$\delta_{1,b} = \boldsymbol{\zeta}_{1,m+1}^{\mathrm{T}}\boldsymbol{b}_{\mathrm{KDOP}}^{(m,n)} . \tag{4.101}$$

同时，得到系数向量 $\boldsymbol{a}_{\mathrm{KDOP}}^{(m+1,n)}$ 的增量形式：

$$\boldsymbol{a}_{\mathrm{KDOP}}^{(m+1,n)} = \boldsymbol{G}^{(m+1,n)}\boldsymbol{C}_{m+1,n}^{\mathrm{T}}\boldsymbol{K}_{1,m+1}^{+}\boldsymbol{g}_{m+1}^{(1)} = \begin{cases} \boldsymbol{a}_{\mathrm{KDOP}}^{(m,n)}, & \lambda_{1,m+1} = 0, \\ \boldsymbol{a}_{\mathrm{KDOP}}^{(m,n)} + \delta_{1,a}\boldsymbol{\theta}_{m+1}, & \lambda_{1,m+1} > 0, \end{cases} \tag{4.102}$$

其中

$$\delta_{1,a} = \boldsymbol{\zeta}_{1,m+1}^{\mathrm{T}}\boldsymbol{a}^{(m,n)} + \frac{g_{m+1}^{(1)} - \boldsymbol{\tau}_{1,m+1}^{\mathrm{T}}\boldsymbol{g}_m^{(1)}}{\lambda_{1,m+1}^{1/2}}\left(1 + \boldsymbol{\zeta}_{1,m+1}^{\mathrm{T}}\boldsymbol{\varepsilon}_{m+1}\right). \tag{4.103}$$

因此得到如下定理[437]。

**定理 4.9（KDOP 的增量学习 I：异类别增量）** 相对于类别（1）的增量数据，类别（2）之 KDOP 的增量形式为

$$f_{\mathrm{KDOP}}^{(m+1,n)}(\boldsymbol{x}) = \begin{cases} f_{\mathrm{KDOP}}^{(m,n)}(\boldsymbol{x}), & \lambda_{1,m+1} = 0, \\ f_{\mathrm{KDOP}}^{(m,n)}(\boldsymbol{x}) + \Delta_1(\boldsymbol{x}), & \lambda_{1,m+1} > 0, \end{cases} \tag{4.104}$$

其中

$$\Delta_1(\boldsymbol{x}) = \delta_1 \sum_{i=1}^{n} \boldsymbol{\theta}_{m+1}(i)\, k(\boldsymbol{x}, \boldsymbol{x}_i^{(2)}) , \tag{4.105}$$

而

$$\delta_1 = \delta_{1,b} - \delta_{1,a} . \tag{4.106}$$

据定理 4.9 的增量形式，可以对模型进行稀疏化。

事实上，当式（4.90）中的 $\lambda_{1,m+1} = 0$ 时，类别（1）的增量数据因对类别（2）之 KDOP 而言无新息而可舍去。而当 $\lambda_{1,m+1} > 0$ 时，由于

$$f_{\mathrm{KDOP}}^{(m+1,n)}(\boldsymbol{x})-f_{\mathrm{KDOP}}^{(m,n)}(\boldsymbol{x})=\Delta_1(\boldsymbol{x}), \tag{4.107}$$

所以，可以定义新息增量

$$\gamma_{1,m+1}=\frac{\left\|f_{\mathrm{KDOP}}^{(m+1,n)}(\boldsymbol{x})-f_{\mathrm{KDOP}}^{(m,n)}(\boldsymbol{x})\right\|}{\left\|f_{\mathrm{KDOP}}^{(m,n)}(\boldsymbol{x})\right\|}, \tag{4.108}$$

并据式（4.82）、式（4.105）、式（4.106）和式（4.107）可得

$$\gamma_{1,m+1}=|\delta_1|\left\{\frac{\boldsymbol{\theta}_{m+1}^{\mathrm{T}}\boldsymbol{K}_{2,n}\boldsymbol{\theta}_{m+1}}{\left[\boldsymbol{b}_{\mathrm{KDOP}}^{(m,n)}-\boldsymbol{a}_{\mathrm{KDOP}}^{(m,n)}\right]^{\mathrm{T}}\boldsymbol{K}_{2,n}\left[\boldsymbol{b}_{\mathrm{KDOP}}^{(m,n)}-\boldsymbol{a}_{\mathrm{KDOP}}^{(m,n)}\right]}\right\}^{1/2}. \tag{4.109}$$

一旦新息增量 $\gamma_{1,m+1}$ 低于预定的阈值（如 0.001），也可以舍去类别（1）的增量数据，最终达到 KDOP 稀疏化的目的。

2. KDOP 的增量学习与稀疏化：对第（2）类样本

类似地，对第（2）类样本，考虑在训练集 $G^{(2)}=\{\boldsymbol{x}_j^{(2)},g_j^{(2)}\}_{j=1}^{n}$ 的基础上增加 $\{\boldsymbol{x}_{n+1}^{(2)},g_{n+1}^{(2)}\}$ 时的影响，即 $f_{\mathrm{KDOP}}^{(m,n+1)}(\boldsymbol{x})$ 与 $f_{\mathrm{KDOP}}^{(m,n)}(\boldsymbol{x})$ 的关系，体现为系数向量是否修正和如何修正。增量学习旨在将 $f_{\mathrm{KDOP}}^{(m,n+1)}(\boldsymbol{x})$ 表示为用 $\{\boldsymbol{x}_{n+1}^{(2)},g_{n+1}^{(2)}\}$ 对 $f_{\mathrm{KDOP}}^{(m,n)}(\boldsymbol{x})$ 加以修正的形式。

首先，由式（4.62）容易得到标签向量 $\boldsymbol{g}_{n+1}^{(2)}$ 的增量形式：

$$\boldsymbol{g}_{n+1}^{(2)}=\mathcal{A}_{2,n+1}f_0(\boldsymbol{x})=\sum_{i=1}^{n+1}<f_0(\boldsymbol{x}),k(\boldsymbol{x},\boldsymbol{x}_i^{(2)})>\boldsymbol{e}_i=\left[\{\boldsymbol{g}_n^{(2)}\}^{\mathrm{T}},g_{n+1}^{(2)}\right]^{\mathrm{T}}. \tag{4.110}$$

而参照式（4.86）和式（4.87）可得 $\boldsymbol{C}_{m,n+1}$ 的增量形式：

$$\boldsymbol{C}_{m,n+1}=\left[\boldsymbol{c}_{m,1},\boldsymbol{c}_{m,2},\cdots,\boldsymbol{c}_{m,n+1}\right]=\left[\boldsymbol{C}_{m,n},\boldsymbol{c}_{m,n+1}\right], \tag{4.111}$$

因此有

$$\boldsymbol{C}_{m,n+1}^{\mathrm{T}}K_{1,m}^{+}\boldsymbol{C}_{m,n+1}=\begin{bmatrix}\boldsymbol{C}_{m,n}^{\mathrm{T}}K_{1,m}^{+}\boldsymbol{C}_{m,n} & \boldsymbol{C}_{m,n}^{\mathrm{T}}K_{1,m}^{+}\boldsymbol{c}_{m,n+1}\\ \boldsymbol{c}_{m,n+1}^{\mathrm{T}}K_{1,m}^{+}\boldsymbol{C}_{m,n} & \boldsymbol{c}_{m,n+1}^{\mathrm{T}}K_{1,m}^{+}\boldsymbol{c}_{m,n+1}\end{bmatrix}. \tag{4.112}$$

其次，显然 $\boldsymbol{K}_{2,n+1}$ 有增量形式[361]：

$$\boldsymbol{K}_{2,n+1}=\begin{bmatrix}\boldsymbol{K}_{2,n} & \boldsymbol{\xi}_{2,n+1}\\ \boldsymbol{\xi}_{2,n+1}^{\mathrm{T}} & \lambda_{2,n+1}\end{bmatrix}, \tag{4.113}$$

其中

$$\lambda_{2,n+1}=k(\boldsymbol{x}_{n+1}^{(2)},\boldsymbol{x}_{n+1}^{(2)}), \tag{4.114}$$

$$\boldsymbol{\xi}_{2,n+1}\equiv\mathcal{A}_{2,n}k(\boldsymbol{x},\boldsymbol{x}_{n+1}^{(2)})=\left[k(\boldsymbol{x}_{n+1}^{(2)},\boldsymbol{x}_1^{(2)}),k(\boldsymbol{x}_{n+1}^{(2)},\boldsymbol{x}_2^{(2)}),\cdots,k(\boldsymbol{x}_{n+1}^{(2)},\boldsymbol{x}_n^{(2)})\right]^{\mathrm{T}}. \tag{4.115}$$

结合式（4.68）、式（4.112）、式（4.113）可得

$$\boldsymbol{K}_{1,\perp,m,n+1}=\boldsymbol{K}_{2,n+1}-\boldsymbol{C}_{m,n+1}^{\mathrm{T}}\boldsymbol{K}_{1,m}^{+}\boldsymbol{C}_{m,n+1}=\begin{bmatrix}\boldsymbol{K}_{1,\perp,m,n} & \boldsymbol{\zeta}_{2,n+1}\\ \boldsymbol{\zeta}_{2,n+1}^{\mathrm{T}} & \sigma_{2,n+1}\end{bmatrix},\tag{4.116}$$

其中

$$\sigma_{2,n+1}=\lambda_{2,n+1}-\boldsymbol{c}_{m,n+1}^{\mathrm{T}}\boldsymbol{K}_{1,m}^{+}\boldsymbol{c}_{m,n+1},\quad \boldsymbol{\zeta}_{2,n+1}=\xi_{2,n+1}-\boldsymbol{C}_{m,n}^{\mathrm{T}}\boldsymbol{K}_{1,m}^{+}\boldsymbol{c}_{m,n+1}.\tag{4.117}$$

再次，由式（4.115）可知$\xi_{2,n+1}\in R(\mathcal{A}_{2,n})\equiv R(\boldsymbol{K}_{2,n})$，再结合式（4.96）可知

$$\boldsymbol{\zeta}_{2,n+1}\in R(\boldsymbol{K}_{2,n})=R(\boldsymbol{K}_{1,\perp,m,n}).\tag{4.118}$$

注意到$\boldsymbol{c}_{m,n+1}=\mathcal{A}_{1,m}k(\boldsymbol{x},\boldsymbol{x}_{n+1}^{(2)})$，$\boldsymbol{K}_{1,m}=\mathcal{A}_{1,m}\mathcal{A}_{1,m}^{*}$，而$\mathcal{A}_{1,m}^{*}\left[\mathcal{A}_{1,m}\mathcal{A}_{1,m}^{*}\right]^{+}=\mathcal{A}_{1,m}^{+}$，因此对式（4.117）中的$\sigma_{2,n+1}$有

$$\begin{aligned}\sigma_{2,n+1}&=k(\boldsymbol{x}_{n+1}^{(2)},\boldsymbol{x}_{n+1}^{(2)})-<\left[\mathcal{A}_{1,m}\mathcal{A}_{1,m}^{*}\right]^{+}\mathcal{A}_{1,m}k(\boldsymbol{x},\boldsymbol{x}_{n+1}^{(2)}),\mathcal{A}_{1,m}k(\boldsymbol{x},\boldsymbol{x}_{n+1}^{(2)})>\\&=<k(\boldsymbol{x},\boldsymbol{x}_{n+1}^{(2)}),k(\boldsymbol{x},\boldsymbol{x}_{n+1}^{(2)})>-<\mathcal{A}_{1,m}^{+}\mathcal{A}_{1,m}k(\boldsymbol{x},\boldsymbol{x}_{n+1}^{(2)}),k(\boldsymbol{x},\boldsymbol{x}_{n+1}^{(2)})>\\&=<\mathcal{P}_{N(\mathcal{A}_{1,m})}k(\boldsymbol{x},\boldsymbol{x}_{n+1}^{(2)}),k(\boldsymbol{x},\boldsymbol{x}_{n+1}^{(2)})>\end{aligned}$$

由于$\mathcal{P}_{N(\mathcal{A}_{1,m})}$是非负定的，所以有

$$\sigma_{2,n+1}=<\mathcal{P}_{N(\mathcal{A}_{1,m})}k(\boldsymbol{x},\boldsymbol{x}_{n+1}^{(2)}),k(\boldsymbol{x},\boldsymbol{x}_{n+1}^{(2)})>\geqslant 0.\tag{4.119}$$

结合式（4.116）、式（4.118）、式（4.119）和文献[434]的定理 3 可得$\boldsymbol{G}^{(m,n+1)}$的增量形式：

$$\boldsymbol{G}^{(m,n+1)}=\boldsymbol{K}_{1,\perp,m,n+1}^{+}=\begin{cases}\begin{bmatrix}\boldsymbol{G}^{(m,n)} & \boldsymbol{0}_{n\times 1}\\ \boldsymbol{0}_{1\times n} & 0\end{bmatrix}, & \sigma_{2,n+1}=0,\\ \begin{bmatrix}\boldsymbol{G}^{(m,n)}+\dfrac{\boldsymbol{\tau}_{2,n+1}\boldsymbol{\tau}_{2,n+1}^{\mathrm{T}}}{\sigma_{2,n+1}} & -\dfrac{\boldsymbol{\tau}_{2,n+1}}{\sigma_{2,n+1}}\\ -\dfrac{\boldsymbol{\tau}_{2,n+1}^{\mathrm{T}}}{\sigma_{2,n+1}} & \dfrac{1}{\sigma_{2,n+1}}\end{bmatrix}, & \sigma_{2,n+1}>0,\end{cases}\tag{4.120}$$

其中

$$\boldsymbol{\tau}_{2,n+1}=\boldsymbol{G}^{(m,n)}\boldsymbol{\zeta}_{2,n+1}.\tag{4.121}$$

最后，得到系数向量$\boldsymbol{b}_{\mathrm{KDOP}}^{(m,n+1)}$的增量形式：

$$\boldsymbol{b}_{\mathrm{KDOP}}^{(m,n+1)}=\boldsymbol{G}^{(m,n+1)}\boldsymbol{g}_{n+1}^{(2)}=\begin{cases}\begin{bmatrix}\boldsymbol{b}_{\mathrm{KDOP}}^{(m,n)}\\ 0\end{bmatrix}, & \sigma_{2,n+1}=0,\\ \begin{bmatrix}\boldsymbol{b}_{\mathrm{KDOP}}^{(m,n)}-\delta_{2,b}\boldsymbol{\tau}_{2,n+1}\\ \delta_{2,b}\end{bmatrix}, & \sigma_{2,n+1}>0,\end{cases}\tag{4.122}$$

其中

$$\delta_{2,b}=\frac{g_{n+1}^{(2)}-\boldsymbol{\tau}_{2,n+1}^{\mathrm{T}}\boldsymbol{g}_{n}^{(2)}}{\sigma_{2,n+1}}.\tag{4.123}$$

同时得到系数向量 $\boldsymbol{a}_{\text{KDOP}}^{(m,n+1)}$ 的增量形式：

$$\boldsymbol{a}_{\text{KDOP}}^{(m,n+1)}=\boldsymbol{G}^{(m,n+1)}\boldsymbol{C}_{m,n+1}^{\text{T}}\boldsymbol{K}_{1,m}^{+}\boldsymbol{g}_{m}^{(1)}=\begin{cases}\begin{bmatrix}\boldsymbol{a}_{\text{KDOP}}^{(m,n)}\\0\end{bmatrix}, & \sigma_{2,n+1}=0,\\ \begin{bmatrix}\boldsymbol{a}_{\text{KDOP}}^{(m,n)}-\delta_{2,a}\boldsymbol{\tau}_{2,n+1}\\ \delta_{2,a}\end{bmatrix}, & \sigma_{2,n+1}>0,\end{cases} \tag{4.124}$$

其中

$$\delta_{2,a}=\frac{\left(\boldsymbol{c}_{m,n+1}-\boldsymbol{C}_{m,n}\boldsymbol{\tau}_{2,n+1}\right)^{\text{T}}K_{1,m}^{+}\boldsymbol{g}_{m}^{(1)}}{\sigma_{2,n+1}}. \tag{4.125}$$

因此，得到如下定理[437]。

**定理 4.10（KDOP 的增量学习 II：同类别增量）** 相对于类别（2）的增量数据，类别（2）之 KDOP 的增量形式为

$$f_{\text{KDOP}}^{(m,n+1)}(\boldsymbol{x})=\begin{cases}f_{\text{KDOP}}^{(m,n)}(\boldsymbol{x}), & \sigma_{2,n+1}=0,\\ f_{\text{KDOP}}^{(m,n)}(\boldsymbol{x})+\Delta_2(\boldsymbol{x}), & \sigma_{2,n+1}>0,\end{cases} \tag{4.126}$$

其中

$$\Delta_2(\boldsymbol{x})=\delta_2\left[k(\boldsymbol{x},\boldsymbol{x}_{n+1}^{(2)})-\sum_{i=1}^{n}\boldsymbol{\tau}_{2,n+1}(i)k(\boldsymbol{x},\boldsymbol{x}_{i}^{(2)})\right],\quad \delta_2=\delta_{2,b}-\delta_{2,a}. \tag{4.127}$$

据定理 4.10 的增量形式，可以对模型进行稀疏化。

事实上，当式（4.117）中的 $\sigma_{2,n+1}=0$ 时，类别（2）的增量数据因对本类 KDOP 而言无新息而可舍去；而当 $\sigma_{2,n+1}>0$ 时，由于

$$f_{\text{KDOP}}^{(m,n+1)}(\boldsymbol{x})-f_{\text{KDOP}}^{(m,n)}(\boldsymbol{x})=\Delta_2(\boldsymbol{x}), \tag{4.128}$$

所以，可以定义新息增量

$$\gamma_{2,n+1}=\frac{\left\|f_{\text{KDOP}}^{(m,n+1)}(\boldsymbol{x})-f_{\text{KDOP}}^{(m,n)}(\boldsymbol{x})\right\|}{\left\|f_{\text{KDOP}}^{(m,n)}(\boldsymbol{x})\right\|}, \tag{4.129}$$

并据式（4.117）、式（4.121）和式（4.127）可得

$$\gamma_{2,n+1}=\left|\delta_2\right|\left\{\frac{\lambda_{2,n+1}-2\boldsymbol{\tau}_{2,n+1}^{\text{T}}\boldsymbol{\xi}_{2,n+1}+\boldsymbol{\tau}_{2,n+1}^{\text{T}}\boldsymbol{K}_{2,n}\boldsymbol{\tau}_{2,n+1}}{\left[\boldsymbol{b}_{\text{KDOP}}^{(m,n)}-\boldsymbol{a}_{\text{KDOP}}^{(m,n)}\right]^{\text{T}}\boldsymbol{K}_{2,n}\left[\boldsymbol{b}_{\text{KDOP}}^{(m,n)}-\boldsymbol{a}_{\text{KDOP}}^{(m,n)}\right]}\right\}^{1/2}. \tag{4.130}$$

一旦新息增量 $\gamma_{1,n+1}$ 低于预定的阈值（如 0.001），也可以舍去类别（2）的增量数据，最终达到 KDOP 稀疏化的目的。

3. 数值计算实例[438]

**例 4.1** 以基于双通道采样的带限信号动态恢复为例，说明增量 KDOP 学习的有效性。

对于例 3.1 中带宽为 π/2 的带限信号 $f_0(x) \in S = \mathrm{span}\{\psi_1(x), \psi_2(x), \cdots, \psi_{15}(x)\}$，其中 $\psi_n(x) = k(x, x'_n)$，$\{x'_n\}_{n=1}^{15} = \{2n\}_{n=-7}^{7}$。

对 $f_0(x)$进行双通道采样，对应的采样函数分别对应子空间 $S_1 = \mathrm{span}\{\psi_1(x), \psi_3(x), \cdots, \psi_{15}(x)\}$和 $S_2 = \mathrm{span}\{\psi_2(x), \psi_4(x), \cdots, \psi_{14}(x)\}$。

设采样中信噪比为 5dB，图 4.7 为一原始信号（$f_0$），以及用双通道采样（$\boldsymbol{g}^{(1)}$、$\boldsymbol{g}^{(2)}$分别表示通道一、通道二）进行信号恢复的结果（$f^{(m,n)}$）。可见，用不断增加的样本对历史恢复结果不断地更新，原信号恢复精度不断增加。若信噪比更高，则恢复得更精确。

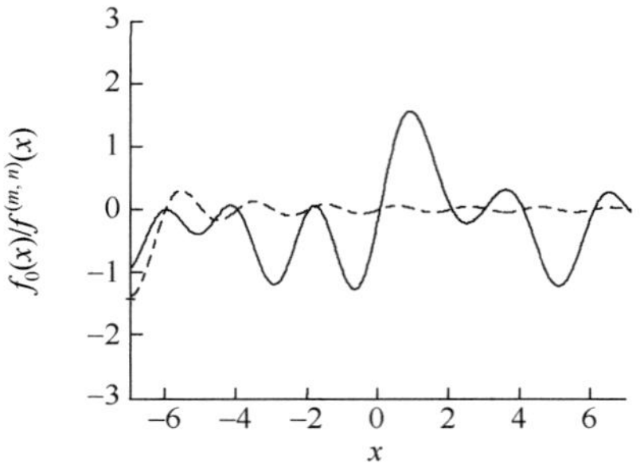

(a) $m = 1, n = 0$

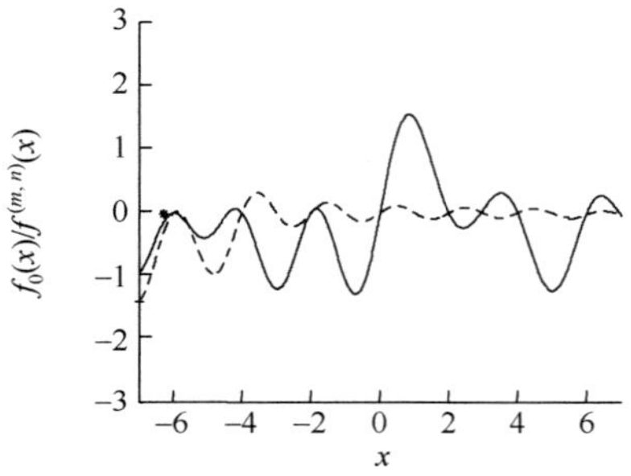

(b) $m = 1, n = 1$

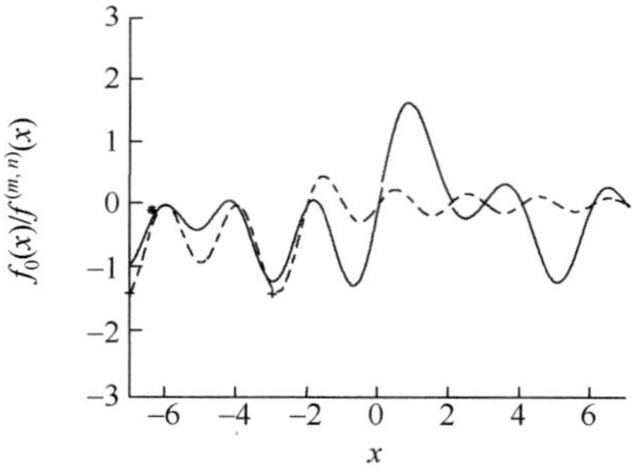

(c) $m = 2, n = 1$

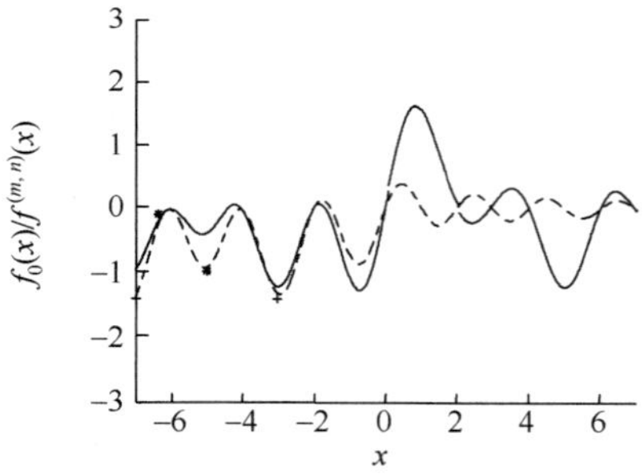

(d) $m = 2, n = 2$

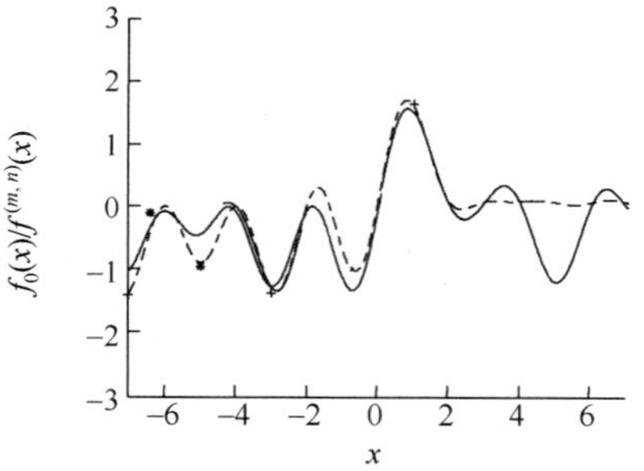

(e) $m = 3, n = 2$

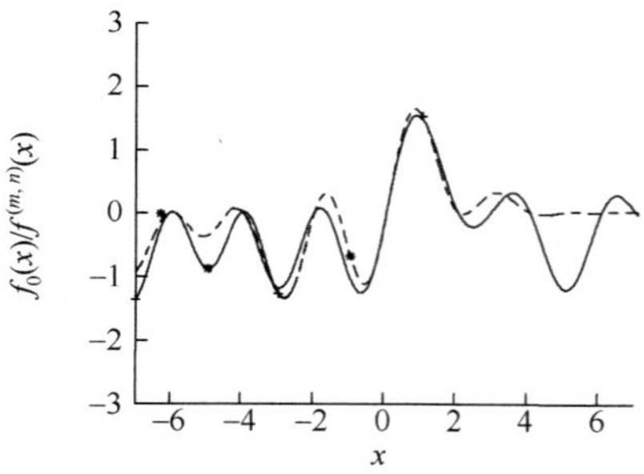

(f) $m = 3, n = 3$

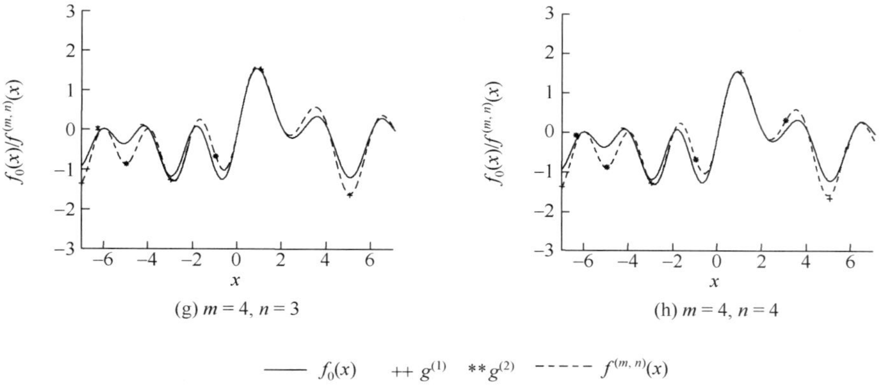


图 4.7　基于双通道采样的信号动态恢复[438]

## 4.5　本 章 小 结

本章从类别区分性角度讨论鉴别型投影学习。在介绍 KND 的投影学习准则、一般形式、增量形式和斜投影扩展形式的基础上，结合最优泛化要求探讨了 KNRD 投影学习，并针对特殊应用场景需要研究了斜投影核鉴别子（KDOP）的基本形式和增量形式，为模式分类应用奠定了基础。

# 第5章 典型应用

## 5.1 引 言

从前述有关章节（1.3 节、第 3 章、第 4 章）的讨论可以看到，本书之投影学习理论和方法的出发点是将机器学习问题、模式识别问题与传统信息处理问题融入统一的框架——函数逼近与逆问题求解（图 4.3），在寻求学习子（图 4.3 中逆算子 $X$）的基础上给出学习结果（表示子或鉴别子，即图 4.3 中 $f(\boldsymbol{x})$）的基本形式、网络结构和在线学习算法。

本章结合典型案例介绍有关投影学习理论和算法的应用，包括在信号分析处理和模式识别中的应用两个方面。

## 5.2 在信号分析处理中的应用

滤波和复原是信号分析处理中的基本问题，其根本目的在于根据观测或处理结果（图 4.3 中 $\boldsymbol{g}$）有效复原原始信号 $f_0(\boldsymbol{x})$，其中的学习子（图 4.3 中 $X$）就是滤波器，而学习结果 $f(\boldsymbol{x})$就是滤波输出。在信号分析中，希望无论是在观测点还是在非观测点，$f(\boldsymbol{x})$都尽可能逼近 $f_0(\boldsymbol{x})$，这就是最优泛化要求。所以，描述型投影学习理论和算法适于解决这一基本问题。

本节以涉及滤波和复原（最优泛化逼近）问题的几个典型案例介绍描述型投影学习算法的应用，包括基于直方图拟合与分解的图像分割、基于曲面拟合与再采样的图像放大、基于多帧融合的图像超分辨重建和基于曲线拟合的语音端点检测（话音活动检测）与增强等等。

### 5.2.1 基于直方图拟合与分解的图像分割

图像分割是图像分析中的经典课题，旨在将图像分为多个不同的连通区域。图像分割是图像目标识别、图像理解、图像/视频检索等应用的基础。

图像分割本质上是一个数据聚类问题。由于应用场景、所依据特征（灰度起伏/颜色变化/纹理多少等）和分割（聚类）算法不同而产生了很多适用于不同图像的分割算法[439, 440]。其中，基于图像灰度直方图分析的门限法对单目标图像而言是最有效的经典算法[441]，而基于数学形态学处理的分水岭法则对多目标图像更有效[442]。由于对多目标图像而言，基于直方图分析的门限法在确定多个门限方面是一个难点，而分水岭法则因对图像噪声非常敏感而容易出现过分割现象，所以这里借助于描述型投影学习的最优泛化特性，探讨基于直方图拟合与分解的多门限估计问题，以有效分割多目标图像[443]。

### 1. 直方图拟合与分解

直方图拟合与分解，是将图像灰度直方图作为观测结果，即式（1.1）中的 $\boldsymbol{g}$，选择合适的函数系和重构准则去拟合直方图得到灰度值的概率密度曲线 $f(\boldsymbol{x})$（参见图 1.3），再按灰度级递增的方式将 $f(\boldsymbol{x})$之显著成分分离出来，为多门限（门限数、门限值）估计奠定基础。

为简单起见，选择高斯函数系拟合直方图，其宽度参数由文献[444]中的简单方法估计，而中心参数值为各灰度级。利用定理 3.1 的 KNR 学习算法拟合直方图，即将灰度值概率密度曲线$f(\boldsymbol{x})$视为$N$个高斯函数的线性组合结果，组合系数由式(3.12)确定。

在实际实现中，可以采用增量 KNR 学习估计组合系数并稀疏化，得到显著性成分及其系数。这里采用更简单的方法稀疏化：将系数值很小、系数值为负数的成分直接去掉。图 5.1 为一幅原始图像及其灰度直方图的拟合及其分解再组合结果 [443]。

在图 5.1（c）的拟合直方图中共 256 个成分，即灰度级数（$N = 256$）。从图 5.1（d）可以看到，稀疏化基本不改变拟合结果。

### 2. 门限估计与合并

利用直方图拟合和分解结果，对按灰度级递增的每两个相邻显著成分，利用纽曼-皮尔逊准则[445]估计一个门限值（其中的虚警概率可以人为设定，如取为 0.02），由此得到一组候选门限值。例如，由图 5.1（d）将得到 199 个门限值（图 5.2）。

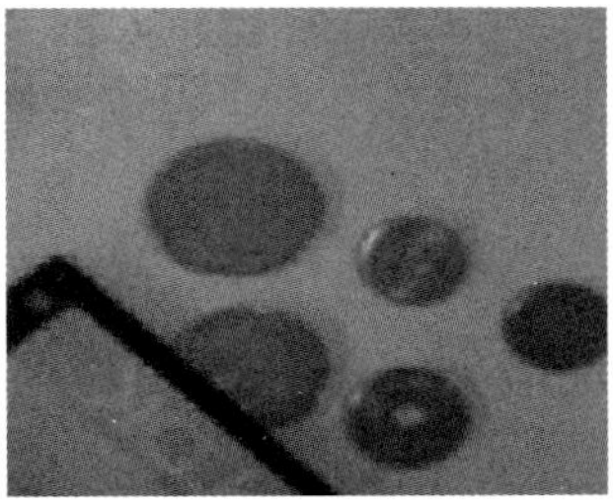

(a) 原始图像

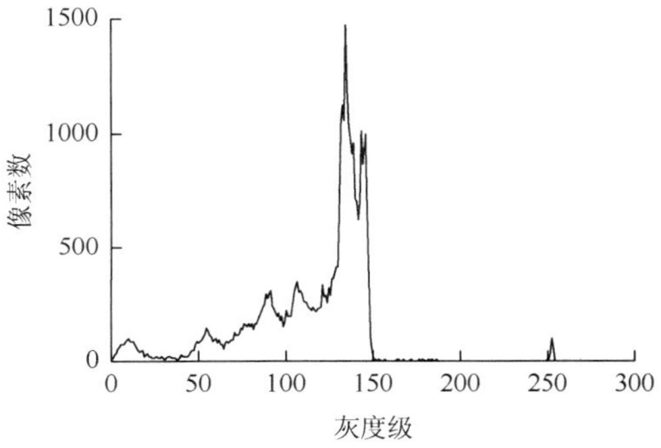


(b) 灰度直方图

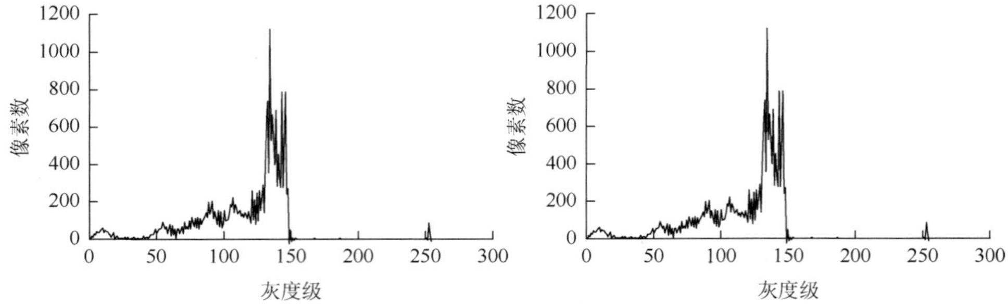


(c) 灰度直方图的KNR拟合（$N=256$）结果　　(d) 拟合曲线的分解（分解为199个显著成分）和组合结果

图 5.1　一幅原始图像及其灰度直方图的拟合及其分解再组合结果[443]

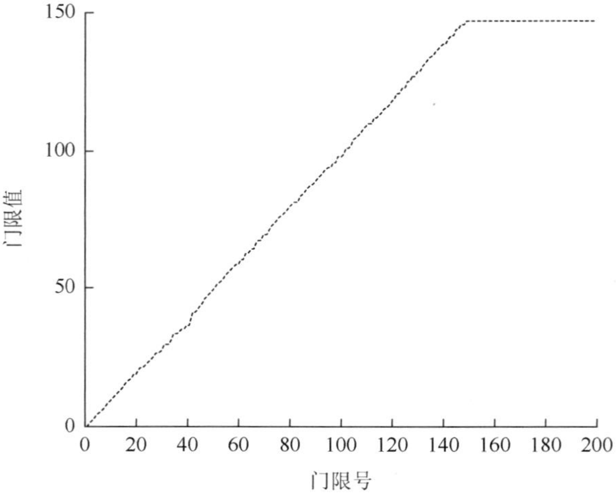


图 5.2　由图 5.1（d）按纽曼-皮尔逊准则得到的候选门限值[443]

由于一些候选门限通常很稠密，需要适当合并以避免过分割现象。

事实上，从单门限问题可以得知[441]，一个有效门限相对于邻近候选门限而言，其值往往具有突变特点。据此，对候选门限序列进行高通滤波可以合并和提取有效门限。在实际应用中，高通滤波器的带宽可以按需要设置（如取 1.76dB 带宽，即将数值低于均值加标准差的预选门限去掉）。另外，对于高通滤波结果中大小差异较小的部分聚集值，可以用其平均值代替，进一步合并有效门限。例如，图 5.2 中的候选门限序列经过高通滤波和聚集值（大小差异低于 16 个灰度级）合并后，得到灰度级分别为 45 和 150 的两个门限。

3. 对比实验结果

对不同场景的多目标图像进行分割，与分水岭算法的分割结果进行对比。图 5.3

列出典型图像分割结果，以及与原图叠加结果。显然，相比分水岭算法，基于直方图 KNR 拟合与分解的算法出现过分割的情况更少。

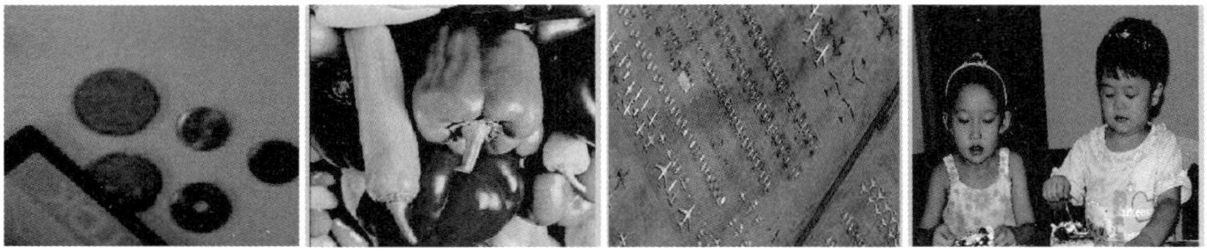

(a) 原图像

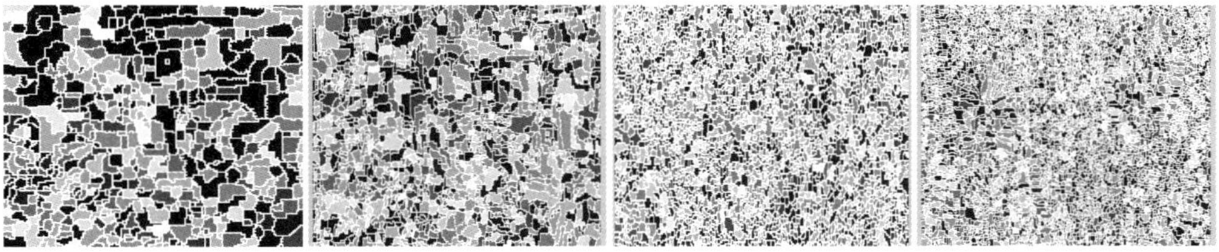

(b) 分水岭法分割结果

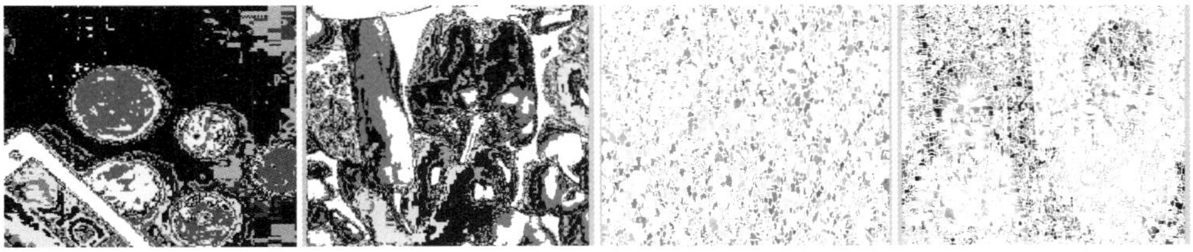

(c) 基于KNR的直方图拟合与分解的分割结果

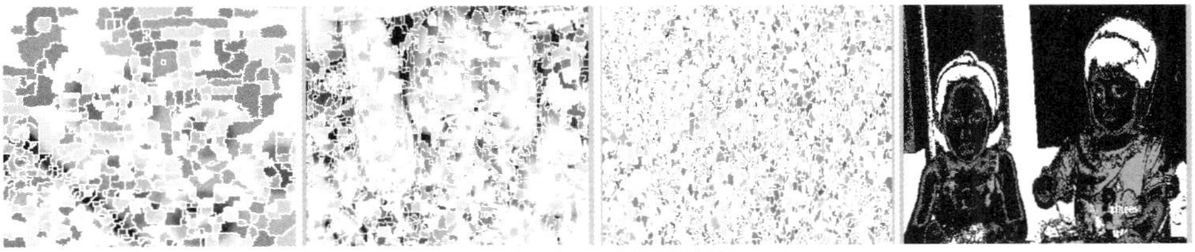

(d) 将(b)与(a)叠加的结果

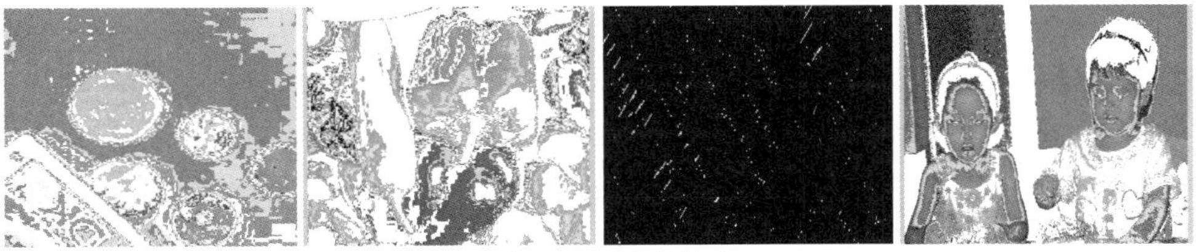

(e) 将(c)与(a)叠加的结果

图 5.3　多目标图像分割[443]

## 5.2.2　基于曲面拟合与再采样的图像放大

图像超分辨重构是信号复原的一个基本问题，旨在利用算法和软件弥补在成像设备或成像条件限制下图像或视频分辨率过低的问题[446]。图像放大又叫单幅图像超分辨重构，是将一幅因分辨率过低而难以有效运用的图像，通过插值等手段提升分辨率、达到有效应用目的的过程。

图像放大的根本任务是利用原始图像邻近像素的相关性产生更多像素点，因此如何填充需要的像素点成为有效放大图像的关键。常用填充像素点的方法有插值和拟合两种，本小节在介绍经典插值放大法的基础上讨论基于 KNR 曲面拟合再采样的放大算法[447, 448]。

1. 经典插值放大法[447]

经典插值法包括最近邻插值法、双线性插值法、双三次插值法，等等。

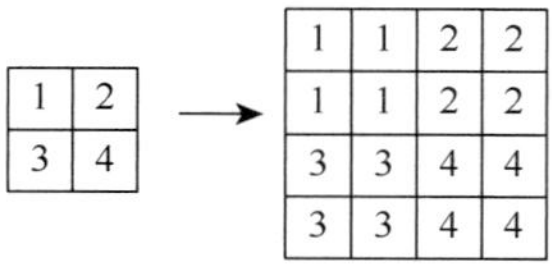


图 5.4　最近邻插值法示意图（放大 2 倍）

1）最近邻插值法

最近邻插值法是将新像素点灰度值取为距离最近的原像素点灰度值。图 5.4 所示为灰度值取 1～4 共 4 个像素点的原始图像经最近邻插值法放大 2 倍的结果。该方法简单、快速，可用于图片浏览等速度要求较高的场合，但有明显的锯齿效应。当放大倍数要求较高时，该方法基本失效。

2）双线性插值法

双线性插值法是用 4 个像素点的灰度值，先沿水平方向线性插值得到新像素点的灰度值，再沿垂直方向由两个新像素点的灰度值插值得到最终像素点的灰度值，示意图如图 5.5 所示，算法如下：

$$\begin{cases} f(b-1,c)=\left[f(b-1,c-1)+f(b-1,c+1)\right]/2, \\ f(b+1,c)=\left[f(b+1,c-1)+f(b+1,c+1)\right]/2, \\ f(b,c)=\left[f(b,c-1)+f(b,c+1)\right]/2, \end{cases} \tag{5.1}$$

式中，$b$ 和 $c$ 为像素点对应矩阵的下标（行号和列号）；$f(b,c)$为用 $f(b-1,c-1)$、$f(b-1,c+1)$、$f(b+1,c-1)$和$f(b+1,c+1)$按式（5.1）插值的结果。

| | | |
|---|---|---|
| $f(b-1,c-1)$ | $f(b-1,c)$ | $f(b-1,c+1)$ |
| $f(b,c-1)$ | $f(b,c)$ | $f(b,c+1)$ |
| $f(b+1,c-1)$ | $f(b+1,c)$ | $f(b+1,c+1)$ |

图 5.5　双线性插值法示意图

双线性插值法与最近邻插值法相比，计算量较大，放大后图像锯齿效应减少，但高频分量受损，图像轮廓模糊。

3）双三次插值法

双三次插值法是用图像中 16 个像素点的灰度值，先沿水平方向每 4 个像素点用三阶多项式拟合插值得到一个新像素点的灰度值，再沿垂直方向用三阶多项式拟合插值得到最终像素点的灰度值，如图 5.6 所示。

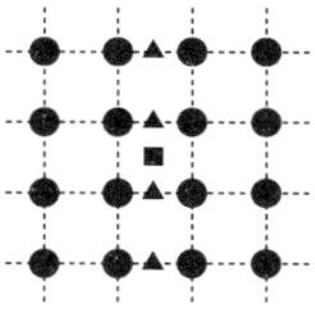

●原像素点；▲水平方向插值点；■垂直方向再插值点

图 5.6　双三次插值法示意图

相对于最近邻插值法和双线性插值法，双三次插值法能有效降低图像锯齿效应和模糊程度。

总的来说，插值方法利用局部邻近像素对图像细节进行估计，忽略了图像全局信息，易带来锯齿效应和细节模糊。

2. 基于 KNR 曲面拟合再采样的放大算法及对比实验[447, 448]

曲面拟合放大算法根据拟合函数和逼近准则用原图像素点拟合出连续曲面，再按所需要的分辨率采样。该算法可较好地抑制锯齿效应和保留细节部分。

这里利用定理 3.1 中的 KNR 算法，通过曲面拟合再采样实现图像任意倍数放大。

为验证 KNR 曲面拟合放大算法的有效性，实验从主观评价和熵、交叉熵、峰值信噪比（peak signal to noise ratio，PSNR）三种客观评价准则进行对比实验。对如图 5.7（a）所示的分辨率为 40×40 的小图像，分别使用三种插值方法和 KNR 曲

(a) 小图像

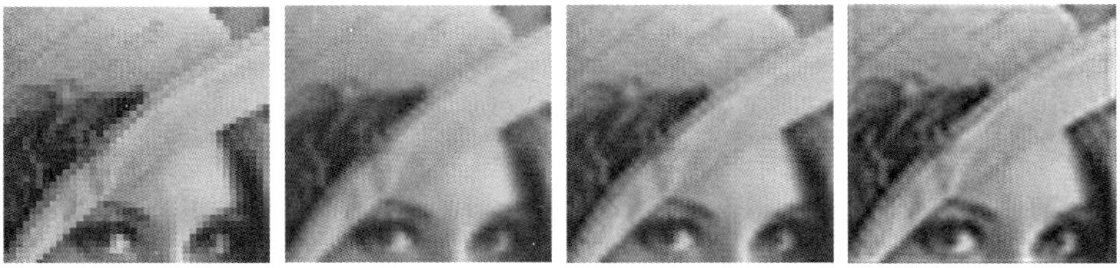

(b) 最近邻插值法　(c) 双线性插值法　(d) 双三次插值法　(e) KNR曲面拟合再采样算法

图 5.7　小图像及经四种不同算法放大 4 倍的结果

面拟合再采样算法放大 4 倍（放大后分辨率为 160×160），结果如图 5.7（b）～图 5.7（e）所示，其客观评价结果如表 5.1 所示。

**表 5.1　小图像不同放大算法性能的客观评价指标（与图 5.7 对应）**

| 评价准则 | 最近邻插值法 | 双线性插值法 | 双三次插值法 | KNR 曲面拟合再采样算法 |
|---|---|---|---|---|
| 熵 | 7.1507 | 7.1934 | 7.2353 | 7.3232 |
| 交叉熵 | 0.0950 | 0.0724 | 0.0564 | 0.0519 |
| PSNR | 22.429 | 23.008 | 23.578 | 23.611 |

从图 5.7 可看到，最近邻插值法的放大图像锯齿效应最明显，双线性插值法的放大图像最模糊且有锯齿效应，双三次插值法的放大图像锯齿和模糊程度相对前两种方法均有明显改善，而 KNR 曲面拟合再采样算法具有更明显的去锯齿效果且细节较清晰。

再从表 5.1 可知，经 KNR 曲面拟合再采样算法放大后，图像的熵较大、交叉熵较小而 PSNR 较大。综合主观和客观因素说明 KNR 曲面拟合再采样算法相对而言是最有效的。

### 5.2.3　基于多帧融合的图像超分辨重建

多帧融合处理是实现图像超分辨重构的主要手段，融合的前提是找到多帧图像的坐标关系。图像间的坐标变化可以是平移、旋转、尺度改变、仿射变换等基本形式，也可以是任意的其他形式。对于任意形式的坐标变化而言，图像配准是估计变化前后坐标关系的基本手段。对于分辨率过低的视频或多帧图像，基于哈里斯（Harris）角点[449]、尺度不变特征变换（scale invariant feature transform，SIFT）特征点[450]等一般特征点的配准算法往往不能实现图像配准，需要重新定义恰当的特征点。

这里定义一种称为特显点（dominant point，DP）的特征点，以其坐标作为低分辨图像配准的依据。对一维信号 $f(x)$，特显点定义为[211, 451]

$$x_k = \underset{x_i \in [a_1, a_2]}{\operatorname{argmin}} \left\{ \sigma \int_{a_1}^{a_2} [f(x) - f(x_i)]^2 \mathrm{d}F(x) + \eta \right\}, \tag{5.2}$$

其右边第一项特显值 $\sigma \int_{a_1}^{a_2} [f(x) - f(x_i)]^2 \mathrm{d}F(x)$，$F(x)$为 $f(x)$在区间$[a_1, a_2]$上的分布函数，$\sigma$ 为对应的方差；$\eta$ 为阈值且 $\eta > 0$。二维信号的特显点可类似定义。

在基于特显点的多帧图像配准和超分辨融合中：首先，根据特显值大小顺序，建立当前图像之特显点与参考图像之特显点的对应关系。其次，将当前图像特显点的坐标向量作为采样点，而参考图像对应特显点的横、纵坐标作为观察值，利用定理3.1中的KNR建立当前图像到参考图像的坐标间非线性映射关系，从而得到当前图像所有像素点在参考图像坐标空间中的新坐标（像素值不变），实现图像配准。然后，利用所有当前图像的新坐标及其对应灰度值，按定理3.3中的KPPLR进行曲面拟合，以期在拟合过程中抑制噪声影响。最后，对拟合曲面按所需分辨率进行再采样，实现任意放大倍数的超分辨重构[451]。

图5.8（a）为录制视频的4帧图像，其中出租车车牌为兴趣块部分，如图5.8（b1）。由于兴趣块图像的分辨率过低，无法提取哈里斯角点等特征点［图5.8（b2）］，而提取的SIFT特征点也因数量过少而不能完成配准［图5.8（b3）］。但提取的特显点数则较多［图5.8（b4）］。特显点配准结果如图5.9（a）所示（为便于观察，显示时这些小图像均按最近邻插值法放大了4倍）。最后，基于KPPLR超分辨重建结果如图5.9（b）所示，而基于三角形的立方插值重建结果如图5.9（c）所示。显然，对车牌处理后的结果（后三位数字“121”）能提供重要的线索，且基于KPPLR超分辨重建结果有更好的视觉效果。

(a) 视频图像帧

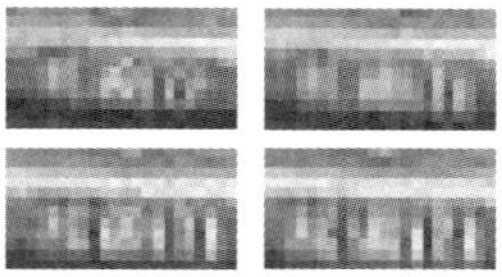

(b1) 兴趣块

(b2) 哈里斯角点

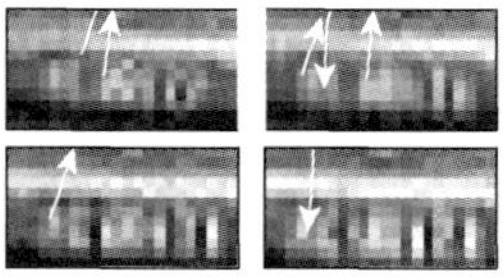

(b3) SIFT特征点

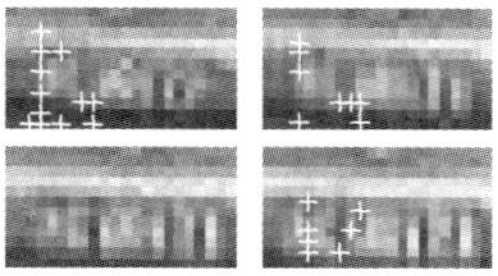

(b4) 特显点

(b) 兴趣块及其三种特征点

图5.8 视频图像帧和兴趣块的三种特征点[211, 451]

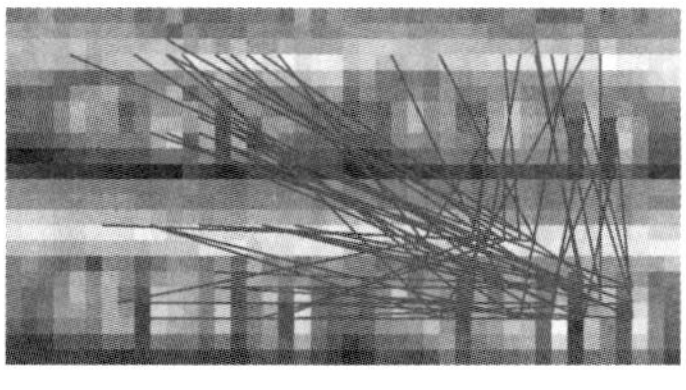

(a) 特显点配准结果

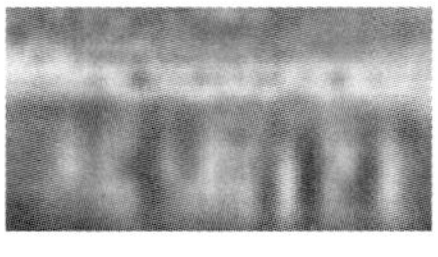

(b) KPPLR超分辨结果

(c) 立方插值超分辨结果

图5.9 基于特显点配准和不同重构算法的超分辨图像[211, 451]

### 5.2.4　基于曲线拟合的语音端点检测与增强

语音端点检测与增强是语音分析应用中一个重要的预处理环节[452]。传统方法主要基于语音信号能量[453]或倒谱等统计特性[454]实现语音端点检测或信号增强。

在我们的研究中，语音端点检测与增强被纳入同一框架——函数逼近与逆问题求解（参见图 4.3），根据输出编码要求或实际输出，实现端点检测和增强[211, 455]。

对含噪语音帧，利用定理 3.1 中的 KNR 进行波形估计（滤波处理）后，计算语音帧的能量、过零率、功率谱密度，构造语音特征向量如下：

$$\boldsymbol{x}_l = w^{(l)}\boldsymbol{F}^{(l)} \quad , \tag{5.3}$$

式中，$\boldsymbol{F}^{(l)}$为第 $l$ 帧语音 $f^{(l)}$的功率谱密度；$w^{(l)}$为增益因子，用过零率 $Z_{\mathrm{CR}}^{(l)}$ 表示为

$$w^{(l)} = \frac{1}{Z_{\mathrm{CR}}^{(l)}}\sum_{i=1}^{n}| f^{(l)}(i)|^2 \quad . \tag{5.4}$$

过零率 $Z_{\mathrm{CR}}^{(l)}$ 的形式为

$$Z_{\mathrm{CR}}^{(l)} = \frac{1}{n}\sum_{i=1}^{n}| \mathrm{sgn}[f^{(l)}(i+1)] - \mathrm{sgn}[f^{(l)}(i)]|^2 \quad , \tag{5.5}$$

其中，sgn 为符号函数。

选取一定数量的非语音帧（明显地只包含背景噪声），提取式（5.3）定义的特征集，训练定理 4.2 中的 KND 作为检测器，完成端点检测。在此基础上，对非平稳噪声方差进行更新，即$\sigma_{(l)}^2 = \eta\sigma_{(l-1)}^2 + (1-\eta)\sigma_l^2$，其中$\sigma_{(0)}^2$为训练帧的平均方差，$\eta$ 为遗忘因子（对语音帧 $\eta = 1$，否则取 $0<\eta<1$ 且$\sigma_l^2$为当前帧方差）。将$\lambda = \sigma_{(l)}^2 \big/ (1+\sigma_{(l)}^2)$作为定理 3.3 中 KPPLR 中的控制参数值，用于语音增强。

图 5.10 所示为用于自动鉴定的一段实录语音样本[小磁带单通道录音、16 位脉冲编码调制（pulse code modulation，PCM）格式、采样率 8kHz]，在不同信噪比下以布莱克曼（Blackman）窗进行加窗分帧（帧长 16ms，重复率 25%）、提取倒谱[454]及式（5.3）所定义特征后，用欧氏距离判据进行端点检测的结果。图中第一列自上至下分别为原始语音，以及信噪比（signal to noise ratio，SNR）分别为 5dB 和 0dB 的含噪语音（加性正态白噪声下）。对应于第一列，第二列为各帧特征与训练帧特征平均值之欧式距离，第三列则为语音端点检测输出结果［粗点线对应于倒谱特征，细实线对应于式（5.3）定义的特征］。显然，所定义特征有更好的端点检测效果（鲁棒性更好）。

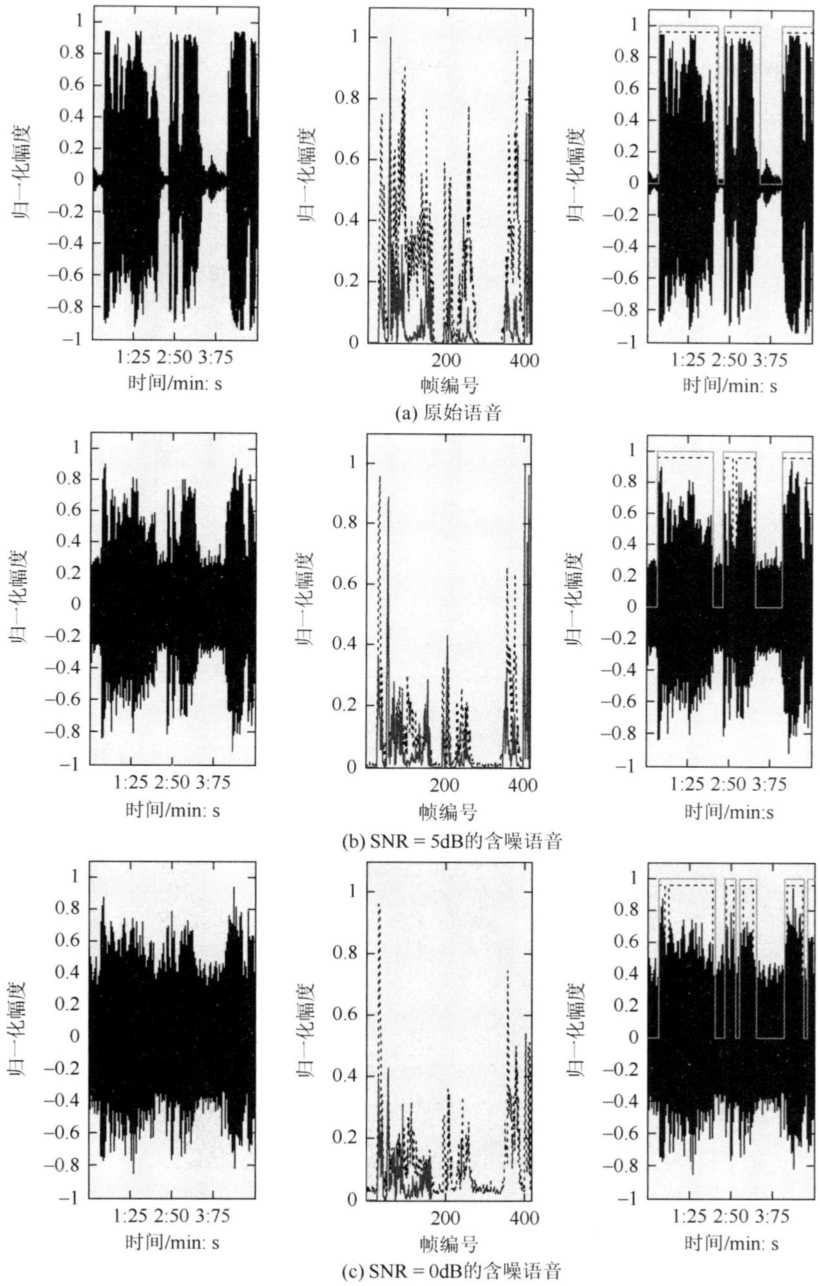


图 5.10 原始语音、含噪语音及其各帧与训练帧之特征的欧式距离（第二列）及端点检测输出（第三列）[211, 455]

粗点线对应于倒谱特征，细实线对应于式（5.3）定义的特征在端点检测的基础上，基于谱减法[456]和 KPPLR 滤波的语音增强结果如图 5.11（a）和图 5.11（b）所示。对比原始语音［图 5.11（c）］可以看到，在波形复原方面 KPPLR 方法优于常用的谱减法。

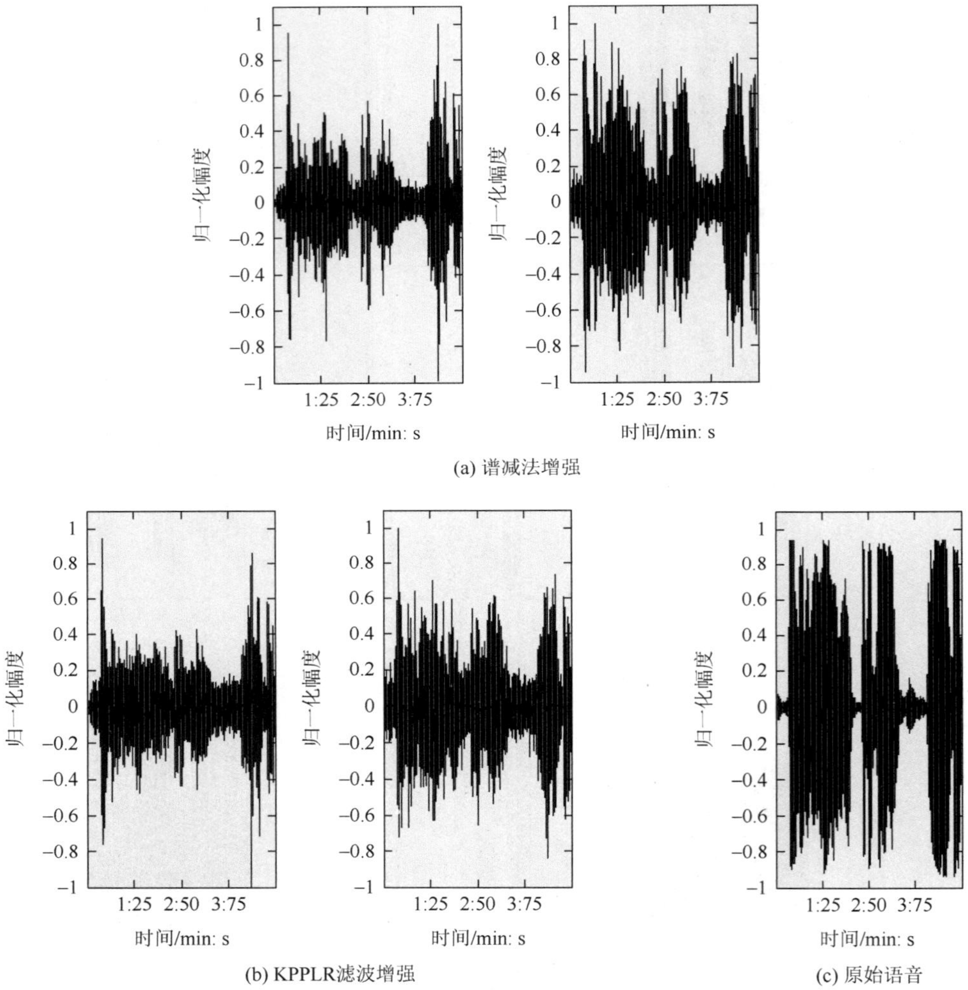


图 5.11　图 5.10 中 SNR 分别为 5dB 和 0dB 的信号增强[211, 455]

## 5.3　在模式识别中的应用

模式识别是人工智能与机器学习的基础任务，也是投影学习理论和算法的主要阵地。特征提取与分类器设计是模式识别的两个重要环节。一般人认为，特征提取在模式识别中起决定作用——只要提取具有足够鉴别能力的模式特征，简单的线性分类器（如欧几里得距离分类器、线性贝叶斯分类器）就足以解决问题。然而，特征提取方法来源于具体问题的应用场景，而且受感测条件限制和不同个体间特征交汇特性影响，在有限测量条件下，模式特征空间中不同个体特征往往是线性不可分的——无论提取何种特征，线性可分只是一种简化或近似，所以，仅仅依靠特征提取难以现实地解决模式识别问题。在现实应用中，恰当设计非线性分类算法往往可以取得满意的识别效果。因此，非线性分类器设计成为模式识别理论及应用研究领域的一个长期课题，受到研究人员的普遍重视。

分类器设计的基本要求是类别区分能力，这正是鉴别型投影学习理论和算法研究的初衷。本节以典型模式识别问题为案例介绍鉴别型投影学习算法的应用，包括手写数字识别、人脸识别、说话人识别、雷达目标识别、视频目标行为识别，等等。

### 5.3.1 手写数字识别

手写字符识别是一个经典的模式识别课题，其中手写数字识别曾在传统邮件分拣应用中扮演过重要角色，也是检验各种模式特征提取手段和分类器有效性的重要手段[457]。为便于对比，这里采用文献[457]中提供的数据库（下载地址为：http://www.ics.uci.edu/～mlearn/MLReposatory.html/），它包含 0～9 这十个数字类别的 2000 个样本（每类 200 个），每个样本是一幅分辨率为 30×48 的图像（典型数据如图 5.12 所示）。

图 5.12 典型手写数字

来自 http://www.ics.uci.edu/～mlearn/MLReposatory.html/

对每个样本，该数据库还提供了以下 6 种典型特征：

（1）字符形状的傅里叶系数（76 维）；

（2）字符轮廓的相关系数（216 维）；

（3）字符图像的 KLT 系数（64 维）；

（4）窗口大小为 2×3 的像素平均结果（240 维）；

（5）字符图像的泽尼克（Zernike）矩（47 维）；

（6）字符图像形态学特征（6 维）。

为检验不同特征下的识别效果，对比 KNR、KND 分类器和文献[457]中所列举的 12 种典型分类器：线性贝叶斯分类器（Bayes Ⅰ）、非线性贝叶斯分类器（Bayes Ⅱ）、最近均值（nearest mean）、最近邻分类器（1-NN）、k 近邻分类器（k-NN）、帕森分类器（Parzen）、费希尔（Fisher）判别准则、决策树（decision tree）、隐层含 20 个节点和 50 个节点的人工神经网络（ANN20 和 ANN50）、线性 SVM 分类器、非线性 SVM 分类器（二阶多项式核）。

在对比实验中，固定 1000 个样本（每类 100 个）作为测试集，从余下 1000 个样本中随机取 500 个（每类 50 个）作为训练集（共十次随机实验），用于估计高斯核宽度[426]的同时统计平均错误分类率（误识率）。对比实验结果如表 5.2 所示[360, 361, 426]，对每一种特征，分类算法效果最好（误识率最小）的结果用加粗字体列出。

**表 5.2　不同特征下各分类器的误识率**

| 特征 | 误识率/% | | | | | | | | | | | | | |
|---|---|---|---|---|---|---|---|---|---|---|---|---|---|---|
| | KNR | KND | Bayes Ⅰ | Bayes Ⅱ | nearest mean | 1-NN | k-NN | Parzen | Fisher | decision tree | ANN-20 | ANN-50 | 线性 SVM | 非线性 SVM |
| （1） | 17.2 | 21.0 | 25.7 | 21.3 | 22.4 | 19.2 | 18.9 | **17.1** | 24.8 | 45.4 | 90.0 | 24.5 | 24.6 | 21.2 |
| （2） | 7.8 | 10.9 | 5.8 | **3.4** | 18.1 | 9.0 | 9.2 | 7.9 | 4.7 | 40.3 | 4.6 | 13.0 | 6.6 | 5.1 |
| （3） | 3.8 | 4.2 | 12.8 | 5.7 | 9.9 | 4.4 | 4.4 | **3.7** | 8.2 | 40.0 | 14.6 | 82.3 | 6.1 | 4.0 |
| （4） | **3.7** | 4.2 | 6.2 | 9.9 | 9.6 | **3.7** | **3.7** | **3.7** | 15.3 | 54.9 | 85.2 | 81.0 | 7.7 | 6.0 |
| （5） | 19.2 | 21.8 | 21.2 | **18.0** | 27.8 | 19.7 | 19.3 | 18.5 | 21.0 | 59.8 | 90.0 | 26.5 | 29.4 | 19.3 |
| （6） | 64.1 | 63.2 | 31.0 | 29.1 | 54.0 | 57.0 | 51.0 | 52.1 | **28.2** | 32.9 | 32.8 | 71.7 | 84.8 | 81.1 |

从表 5.2 可以看到，总体上 KNR 和 KND 的性能优于其他多数分类器。例如，对特征（3）和特征（4），KNR 和 KND 的误识率低于其他多数分类器；而对于 6 种特征而言，KNR 和 KND 与非线性 SVM 分类器性能相近。另外，总体上 KNR 的分类效果略优于 KND。

此外，为考察基于增量学习的稀疏化效果，表 5.3[360, 361]列出自适应 KNR 和自适应 KND 的分类结果，其中稀疏度定义为余留训练样本数与总样本数之比。

表 5.3 不同特征下自适应 KNR 和自适应 KND 的误识率与稀疏度

| 特征 | 自适应 KNR | | 自适应 KND | |
|---|---|---|---|---|
| | 误识率/% | 稀疏度 | 误识率/% | 稀疏度 |
| （1） | 19.6 | 0.87 | 21.9 | 0.84 |
| （2） | 17.6 | 0.79 | 15.2 | 0.75 |
| （3） | 6.8 | 0.80 | 4.2 | 1.00 |
| （4） | 6.1 | 0.75 | 4.2 | 1.00 |
| （5） | 27.0 | 0.76 | 26.5 | 0.50 |
| （6） | 65.8 | 0.60 | 61.9 | 0.05 |

对比表 5.2 和表 5.3 可知，稀疏化对自适应 KND 的分类效果影响不明显，但对自适应 KNR 则越稀疏误识率越高，这是由其描述性特性所决定的（模型越小泛化能力越弱）。

### 5.3.2 人脸识别

机器视觉是人工智能领域的一个核心问题（参见表 1.1 和 1.2.2 小节），其中人脸识别由于在罪犯身份识别、银行及海关监控、证件与持证人核对、自动门卫等安全领域和人机交互中有着广泛应用而备受关注，也是检验各种模式特征提取和分类器算法有效性的重要手段[458-460]。

为验证和比较各种人脸识别算法，学界先后搭建了各种人脸库[459]，其中搭建得最早和被使用最广泛的数据库是剑桥大学计算机实验室的 ORL（Olivetti Research Laboratory）人脸库[461]，其包含 40 个人在不同光照、不同细节（是否戴眼镜或是否留胡须等）、不同表情下的 400 个样本（每人 10 幅正面脸像，分辨率均为 92×112），典型数据如图 5.13 所示。

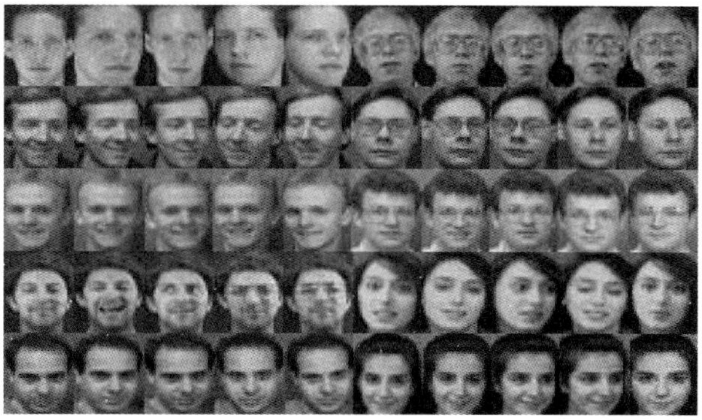

图 5.13 ORL 人脸库中的典型样本[461]

研究表明，对多样本而言，由于 PCA、LDA 等无师学习算法具有发掘数据内在特征和降低数据维数的优势，所以被广泛用于提取人脸的统计特征，对应特征被称为本征脸（eigen faces）、费希尔脸（Fisher faces）等[459, 462]。

由于这些特征提取过程是正交变换，所以最简单的分类可以用欧几里得距离（测试样本特征向量与训练样本特征向量之平均结果之间的距离）最小为判据（简称欧氏距离分类器），以此为基准比较不同特征下不同分类器的识别效果。

这里以 ORL 人脸库进行对比实验。在特征提取前先对每幅图像进行降采样处理（如简单地将图像分为大小为 4×4 的不交叠块，以每块的平均灰度作为一个新像素点的灰度，将分辨率降为 23×28）。与从训练样本预处理图像中直接提取本征脸和费希尔脸相对比，这里先对预处理图像进行快速傅里叶变换（fast Fourier transform，FFT）并取模，再经 PCA 提取特征，得到的结果称为本征谱（eigenspectra）[463]。

特征提取后，根据训练样本的特征进行分类器训练。图 5.14 为基于本征谱特征和 KNRD 分类器的识别过程（基于其他特征或分类器的过程类似）[396]。对基于高斯核的 KNRD 分类器，其宽度参数用训练样本特征按文献[444]之式（8）进行简单估计。

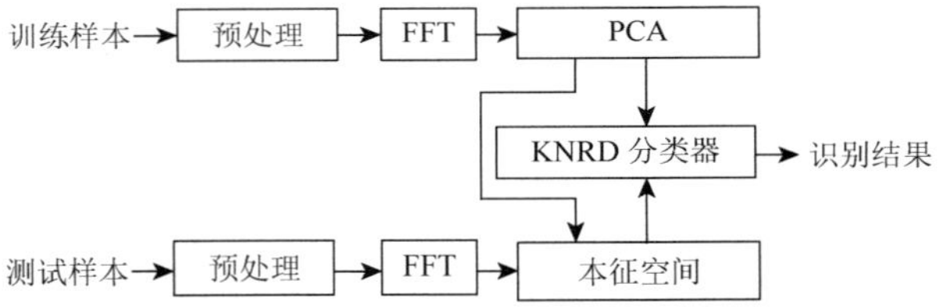


图 5.14　基于本征谱特征和 KNRD 分类器的人脸识别过程[396]

首先，固定分类器（这里为 KNRD）以对比不同特征的识别结果。从 ORL 人脸库中每个类别中随机取 5 个样本（共 200 个样本）进行训练，用余下样本进行测试，统计正确识别率与 KNRD 分类器参数 $\lambda$ 值的关系，重复 10 次实验对正确识别率进行平均的结果如图 5.15 所示。

由图 5.15 可以看到，KNRD 的参数 $\lambda$ 值大于 0.0001 后识别率基本处于稳定状态（与 $\lambda$ 值大小几乎无关），且基于本征谱的识别率比基于本征脸的识别率高出约 5%，可达 97%。

其次，固定特征（这里为本征谱）以对比不同分类器分类效果和不同训练样本数的影响。这里以 KNR、KND、KNRD（其中 $\lambda$ 值取 0.1）和欧氏距离分类器进行对比，并考察每类训练样本数的影响，重复十次随机实验的平均识别率如表 5.4 所示[396]。

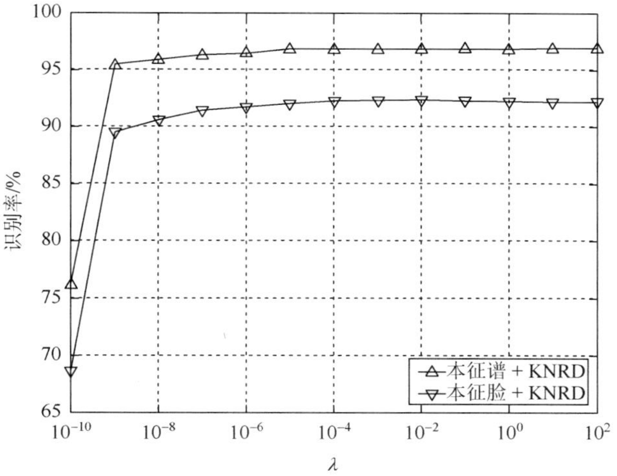


图 5.15 基于本征谱特征和 KNRD 分类器的人脸识别结果[396]

**表 5.4 基于本征谱的不同分类器在不同训练样本数下的识别率**

| 训练样本数 | 识别率/% | | | |
|---|---|---|---|---|
| | KNR | KND | KNRD（$\lambda=0.1$） | 欧氏距离分类器 |
| 2 | 87.6 | 71.7 | 87.4 | 86.7 |
| 3 | 94.2 | 78.8 | 94.2 | 89.9 |
| 4 | 95.0 | 79.2 | 95.1 | 90.7 |
| 5 | 96.8 | 76.0 | 96.9 | 92.4 |
| 6 | 97.8 | 76.1 | 97.9 | 93.9 |
| 7 | 98.4 | 72.8 | 98.4 | 94.5 |
| 8 | 98.0 | 63.9 | 98.0 | 95.0 |

由表 5.4 可以看到，总体上识别率与训练样本数成正变关系，KNR 分类器与 KNRD 分类器相当且优于欧氏距离分类器的效果，而 KND 分类器总体上不如其他分类器。

最后，基于同样的数据库和实验条件，以最佳识别率对比不同识别方法的效果。从图 5.15 和表 5.4 可以看到，基于本征谱和 KNR、KNRD 分类器的识别结果优于其他典型方法，如直接 LDA 法（94.0%）[464]、基于奇异值分解（singular value decomposition，SVD）系数和贝叶斯分类器的方法（92.5%）[465]、基于本征脸和核密度估计的方法（92.5%）[466]、基于鉴别性小波脸和最近特征空间分类的方法（96.1%）[467]，以及单类别 PCA 的方法（95.0%）[468]。

### 5.3.3 说话人识别

与人脸识别类似，说话人识别也是生物特征智能识别中的一个主要课题，在安全领域、人机交互中具有重要的应用价值[469-471]。

语音信号经过预加重、分帧加窗、端点检测等预处理后提取各种特征[211]，其中结合语音产生机制提取的梅尔频率倒谱系数（Mel frequency cepstral coefficients，MFCC）[472]，由于与作为特殊非线性系统的人之听觉系统特点（对不同频率信号有不同的灵敏度，基本呈对数关系）相吻合而在语音识别、声纹鉴定、说话人识别等应用中被广泛采用[469-473]。

在特征建模方面，高斯混合模型（Gaussian mixture model，GMM）作为一种统计分布估计手段，由于能很好地拟合说话人语音特征且便于实时实现而被广泛采用[470, 473]。

作为应用案例，这里基于 MFCC 特征，对比 GMM 法和 KNR 分类法进行说话人识别实验。实验目的主要对比在训练和测试语音之时长不同的情况下，GMM 法和 KNR 分类法的性能。

实验所用语料库是 40 个人的汉语普通话之正常朗读语音的集合，其中每个人发音大约在 4min，采样率为 8kHz，量化位数为 16bit。录制环境比较安静，SNR 比较高。

考虑到每个人语速、发音情况不一样，同样长度的语音片段得到的帧数也不一致，为此对语料库作如下处理：对于每一个人的发音片段，使用相同的预处理得到语音帧，提取 MFCC 特征，形成每个人的特征集。

实验中，提取 16 维 MFCC 特征，选取 16 阶 GMM，而 KNR 分类器中的核函数取高斯核函数，其中期望向量和协方差矩阵由训练样本的样本均值和样本协方差代替。

在训练与测试时间较短的情况下，选择 10 个人的语音进行识别，训练语音片段分别为 50s（1500 帧）、30s（1000 帧）和 10s（330 帧），测试语音片段分别为 30s（1000 帧）、15s（500 帧）和 5s（168 帧），分别考察对应训练和测试时间下的识别率。实验结果分别列入表 5.5[211]。

由表 5.5 可以看出，随着训练语音、测试语音时长的降低，GMM 分类器性能下降很快，而 KNR 分类器性能下降较慢。当训练时长最短（10s）时，KNR 的识别性能优于 GMM。所以，在语料充足的情况下，GMM 的识别性能要高于 KNR，而 KNR 在获得的语音片段较少、类别数较少的情况下占据一定优势。

**表 5.5 不同时长的训练语音下 GMM 与 KNR 的识别率**

| 分类器 | 训练语音时长/s | 识别率/% | | |
|---|---|---|---|---|
| | | 5s① | 15s | 30s |
| GMM | 50 | 42.5 | 70.0 | 70.0 |
| | 30 | 30.0 | 55.0 | 60.0 |
| | 10 | 20.5 | 22.5 | 22.5 |
| KNR | 50 | 70.0 | 80.0 | 85.0 |
| | 30 | 67.5 | 67.5 | 70.0 |
| | 10 | 62.5 | 65.0 | 65.0 |

①为测试语音片段时长。

## 5.3.4 雷达目标识别

自动目标识别（automatic target recognition，ATR）是雷达领域的一个重要课题和机器学习算法的重要用武之地[474-477]，而宽带、超宽带微波技术和现代信号处理技术的快速发展则为基于高分辨成像数据有效地实现 ATR 奠定了坚实的基础[478-486]。其中一维像（又称高分辨雷达距离剖面像）由于内含目标信息丰富、数据冗余相对较少且处理方便，因此成为快速有效实现 ATR 的一种重要手段[479-481]。

为验证投影学习算法在 ATR 中的有效性，这里基于实测一维像和仿真一维像进行目标识别实验。

在实测数据实验中，一维像由某逆合成孔径雷达（inverse synthetic aperture radar，ISAR）对三种飞行中的飞机目标进行实测形成，每种飞机均录取了多段数据，如图 5.16[479, 480]所示。

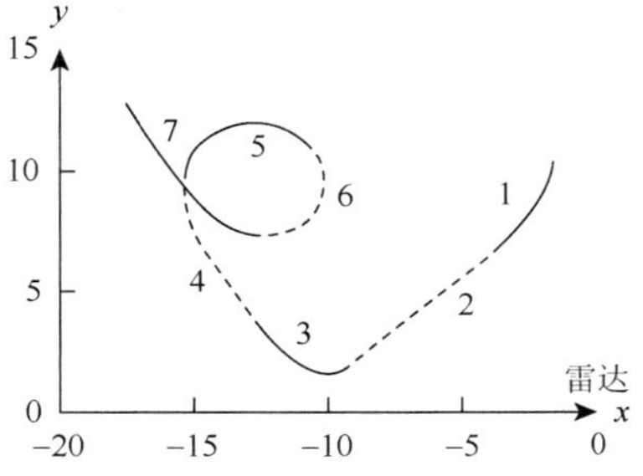


(a) 目标1

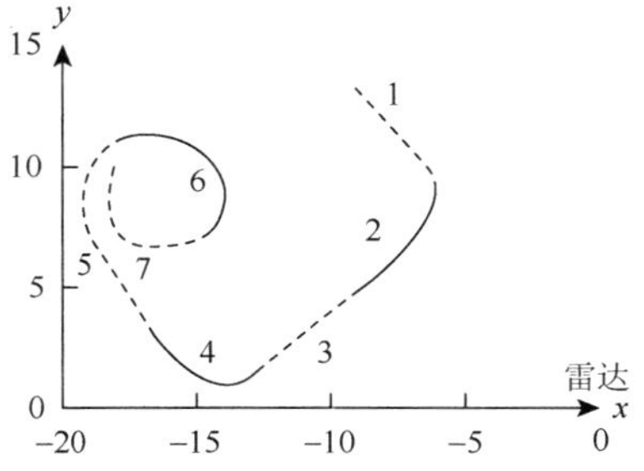


(b) 目标2

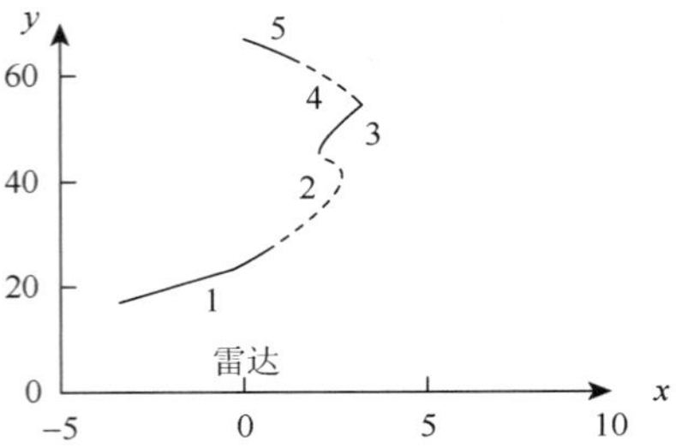


(c) 目标3

图 5.16　三种飞机的平面航迹[479, 480]

三种飞机目标对应的典型一维像如图 5.17 所示。

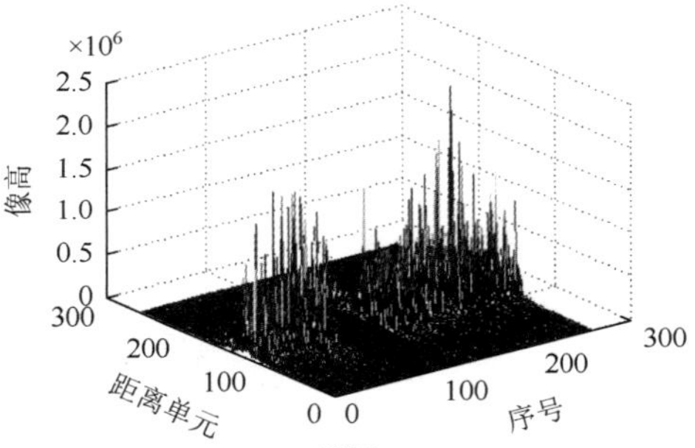


(a) 目标1

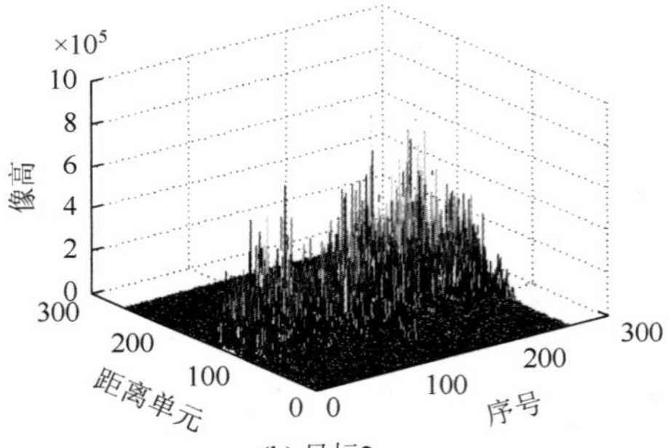


(b) 目标2

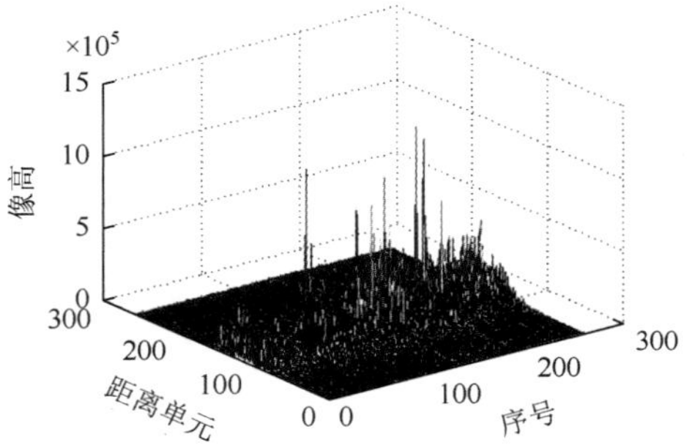


(c) 目标3

图 5.17　三种飞机的典型一维像[480]

利用本征谱特征，基于 KND、KNR、SVM 和 RBF4 种典型分类器的识别率、平均识别率和训练总时间如表 5.6 所示[435]。

**表 5.6　基于不同分类器的三种目标识别率、平均识别率和训练总时间**

| 分类器 | 识别率/% | | | 平均识别率/% | 训练总时间/s |
|---|---|---|---|---|---|
| | 目标 1 | 目标 2 | 目标 3 | | |
| KND | 98.6 | 87.2 | 98.8 | 96.1 | 0.010 |
| KNR | 97.9 | 90.4 | 98.8 | 96.6 | 0.125 |
| SVM | 98.6 | 88.3 | 98.3 | 96.1 | 6.859 |
| RBF | 99.3 | 88.3 | 97.1 | 95.8 | 0.484 |

从表 5.6 可以看到，几种典型分类器的分类效果相当，但 KND 和 KNR 在训练时间上占绝对优势，所以综合性能更优。

在仿真数据实验中，对 E2C、F16、IDF、J8II、Mirage2000 和 Su276 种典型飞机按成像算法生成一维像序列，典型一维像的能量归一化结果如图 5.18[435]所示。

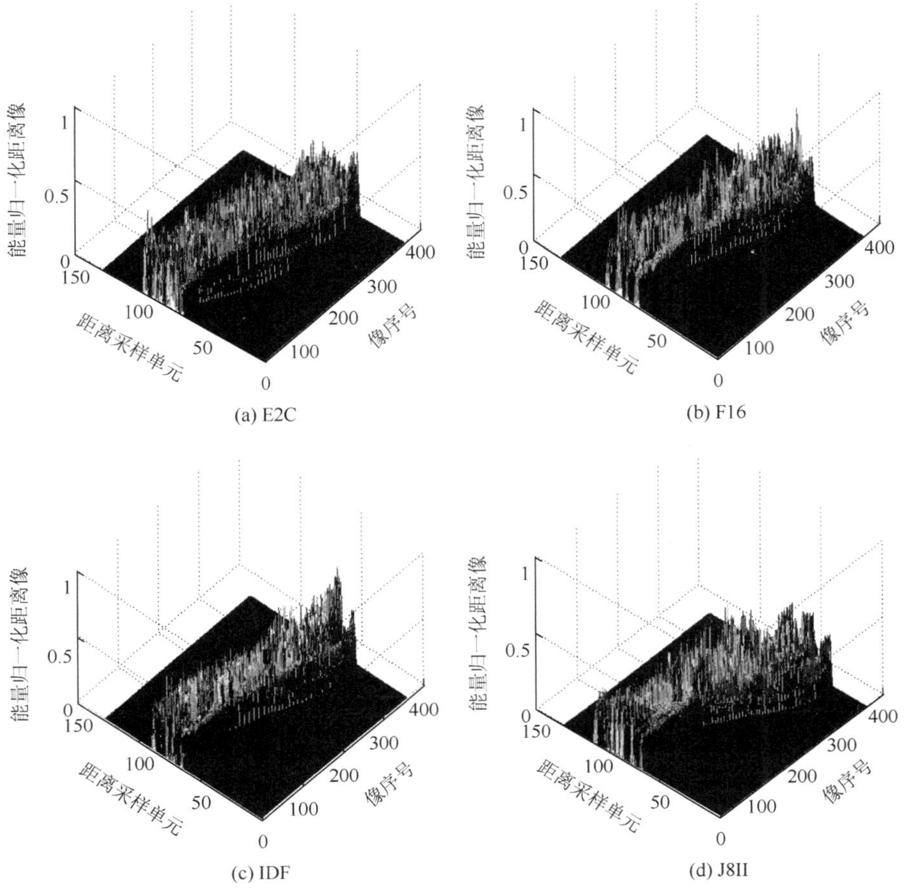

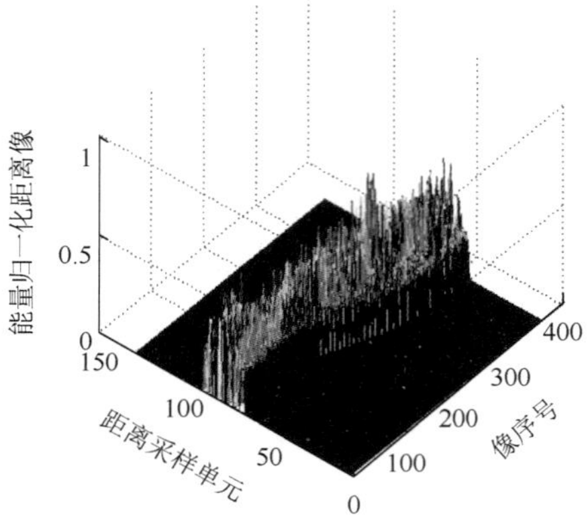

(e) Mirage2000

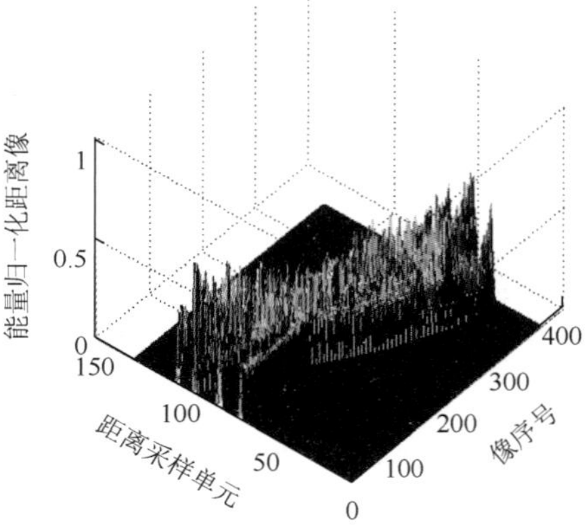

(f) Su27

图 5.18　6 种典型目标的仿真一维像

利用本征谱特征，基于 KND、KNR、SVM 和 RBF 四种典型分类器的识别率、平均识别率和训练总时间如表 5.7[435]所示。

**表 5.7　不同分类器的六种目标识别率、平均识别率和训练总时间**

| 分类器 | 识别率/% | | | | | | 平均识别率/% | 训练总时间/s |
|---|---|---|---|---|---|---|---|---|
| | E2C | F16 | IDF | J8II | Mirage2000 | Su27 | | |
| KND | 100.0 | 95.0 | 97.5 | 67.5 | 97.5 | 100.0 | 92.9 | 0.015 |
| KNR | 100.0 | 95.0 | 97.5 | 70.0 | 95.5 | 100.0 | 92.9 | 0.031 |
| SVM | 100.0 | 95.5 | 95.5 | 70.0 | 92.5 | 100.0 | 92.1 | 1.453 |
| RBF | 100.0 | 95.0 | 97.5 | 65.0 | 95.0 | 100.0 | 92.1 | 0.609 |

从表 5.7 可以看到，几种典型分类器的分类效果相当，但 KND 和 KNR 在训练时间上占绝对优势，所以综合性能更优。

### 5.3.5　视频目标行为识别

视频目标行为识别就是对视频流中目标的行为（包括打、砸、抢、跑、跳、掷、摔等人的行为，以及违规变道、挤占应急车道、山体塌方等物的行为）进行特征提取、描述、动态建模和识别分类，其在基于内容的视频检索和视频摘要、视频监控与公共安全、人机交互、机器人技术、远程医疗、虚拟培训、智能看护、视觉娱乐等领域具有广阔的应用前景[487-489]。

人的行为从难易程度上分为姿态、动作、交互行为和群体行为等多种类别[488, 489]。

视频目标行为分析过程，按层次划分包括视频特征提取（底层）、行为描述及建模和识别分类（中层）、行为语义抽象及推理（高层）等[487, 489]。其中，高层次工作需要结合具体的应用展开，而底层次和中层次工作是整个行为分析过程的基础和关键，也是本小节讨论的主要内容。

从视频中提取的目标行为特征包括光流、关键点/兴趣点轨迹、轮廓/边界流等运动特征，其中稠密光流轨迹（dense optical flow trajectory，DOFT）[490]因对复杂运动模式非常有效而成为一种重要的局部特征。对这些特征的描述、建模和识别分类方法包括基于时空特征体的模板匹配、特征的统计建模等[487-489]。其中特征描述的经典方法有方向梯度直方图（histogram of oriented gradient，HOG）[491]、光流直方图（histogram of optical flow，HOF）[492]、运动边界直方图（motion boundary histogram，MBH）等[493]。为验证投影学习理论和算法的有效性，本小节介绍基于深度 KNR 网络模型的行为识别算法，其核心是对特征描述子进行特征编码与融合。

1. 基于深度网络的特征编码

图 3.1 所示的网络模型，作为 KNR、KPLR、KPPTPLR、KND、KNRD、SVM、RBF 等核非线性网络的共享模型，是描述数据之间复杂非线性关系的经典模型，直接将其扩展为深度网络的工作在机器学习领域备受关注。例如，深度费希尔网络（deep Fisher network，DFN）是将费希尔核网络扩展为深度层结构而得到的网络[494, 495]，卷积核网络（convolutional kernel network，CKN）则通过叠加核非线性映射层、池化层获取样本的高维表达并使用 BP 算法调节非线性层系数而得到深度卷积核网络[496, 497]。此外，还有一些深度较浅的层次表达模型，如借鉴深度网络之中层表达的思想，传统费希尔向量被拓展为堆栈费希尔向量（stacked Fisher vector，SFV）用于视频目标行为表达和行为识别，并在两层之间引入了有监督的特征降维环节[498]。

1）基于深度 KNR 网络的特征编码

核非线性网络的一大优点在于不用显式表达出非线性映射函数，只需设计出映射后样本之间的内积形式（即核函数）即可[328-332]。然而核函数的这种内聚性表达往往限制了核非线性网络的工程应用。为了打破核函数的内聚性，这里探讨一种使用核矩阵显式表达非线性映射函数的编码方法，并通过这种方法将视频的密集轨迹原始特征描述子显式地映射到高维的新特征空间中，实现原始特征的再编码。鉴于定理 3.1 中的 KNR 可以直接处理多分类问题（不用像 SVM 一样需要使用投票法等多次训练鉴别函数），这里基于 KNR 探讨这种行为特征的再编码方式[499]。

对具有再生核 $k$ 的 RKHS $H$，考虑其子空间

$$S = \mathrm{span}\{\psi_1(z), \psi_2(z), \cdots, \psi_{M_1}(z)\} \subset H \,, \tag{5.6}$$

其中，$\psi_i(\boldsymbol{x}) = k(\boldsymbol{z}_i, \boldsymbol{x})$，而 $\boldsymbol{z}_i(i = 1, 2, \cdots, M_1)$是经聚类算法（如 k-means 算法）在所有训练样本特征中生成的 $M_1$ 个簇中心。对于任意输入特征向量 $\boldsymbol{x}$，有

$$\boldsymbol{\gamma}_{\boldsymbol{x}} = (\boldsymbol{K}_1^{+})^{1/2}\boldsymbol{y}, \quad \boldsymbol{y} = \left[k(\boldsymbol{z}_1, \boldsymbol{x}), k(\boldsymbol{z}_2, \boldsymbol{x}), \cdots, k(\boldsymbol{z}_{M_1}, \boldsymbol{x})\right]^{\mathrm{T}}, \tag{5.7}$$

称 $\boldsymbol{\gamma}_{\boldsymbol{x}}$ 为 KNR 编码（KNR coding，KNRc），其中 $\boldsymbol{K}_1$ 为 $\psi_i(\boldsymbol{x}) = k(\boldsymbol{z}_i, \boldsymbol{x})$对应的格拉姆矩阵。

设另一输入特征向量 $\boldsymbol{x}'$经 KNR 编码得到 $\boldsymbol{\gamma}_{\boldsymbol{x}'}$，则参照式（3.11）和式（3.12）可得

$$<\boldsymbol{\gamma}_{\boldsymbol{x}}, \boldsymbol{\gamma}_{\boldsymbol{x}'}>= \boldsymbol{y}^{\mathrm{T}}\boldsymbol{K}^{+}\boldsymbol{y}' = \boldsymbol{y}^{\mathrm{T}}\boldsymbol{K}^{+}\boldsymbol{K}\boldsymbol{K}^{+}\boldsymbol{y}' = (\boldsymbol{K}^{+}\boldsymbol{y})^{\mathrm{T}}\boldsymbol{y}\boldsymbol{y}^{\mathrm{T}}(\boldsymbol{K}^{+}\boldsymbol{y}') =< f(\boldsymbol{x}), f(\boldsymbol{x}') >= k(\boldsymbol{x}, \boldsymbol{x}') \,. \tag{5.8}$$

所以，KNRc 可显式表达非线性映射，且 KNRc 有以下显著优点：一是将 $M$ 维特征映射到 $M_1$ 维特征空间中进行表达，恰当的映射可以更好地表达原始特征；二是在一定程度上可将原特征空间中的非线性关系转换为新特征空间中的线性关系；三是与原 KNR 只包含本类别相似信息相比，可包含不同类别特征向量的相似信息，使得特征更具鉴别性。

2）费希尔向量编码

上述 KNRc 为描述型编码，费希尔向量编码旨在提取鉴别性编码特征。

费希尔向量编码的核心思想是将可处理的、样本维数可变的生成式模型嵌入到鉴别型分类器中。费希尔向量编码起初用于 DNA 信息分类和蛋白质序列分析[500]，后来通过改进以解决大规模图像分类问题 [501]。

设集合 $\boldsymbol{X} = \{\boldsymbol{x}_i\}_{i=1}^{N}$ $(\boldsymbol{x}_i \in R^M)$ 是图像或视频的 $N$ 个特征描述子，用 $u_{\boldsymbol{\theta}}(\boldsymbol{X})$表示生成该描述子的概率密度函数，其中 $\boldsymbol{\theta}$ 是其参数向量。则 $\boldsymbol{X}$ 可描述为梯度向量：

$$\boldsymbol{G}_{\theta}^{X} = \frac{1}{N}\nabla_{\theta}\log u_{\theta}(\boldsymbol{X}) \,. \tag{5.9}$$

梯度反映了参数在生成过程中的贡献程度，而梯度向量的维数仅取决于参数向量 $\boldsymbol{\theta}$ 的维数，与特征描述子的个数 $N$ 无关。使用梯度向量构造的核函数

$$k_F(\boldsymbol{X}, \boldsymbol{Y}) = \left(\boldsymbol{G}_{\theta}^{X}\right)^{\mathrm{T}} \boldsymbol{F}_{\theta}^{-1}\boldsymbol{G}_{\theta}^{Y} \tag{5.10}$$

称为费希尔核，其中 $\boldsymbol{F}_{\boldsymbol{\theta}}$ 是概率密度函数 $u_{\boldsymbol{\theta}}(\boldsymbol{X})$的费希尔信息矩阵，即

$$\boldsymbol{F}_{\theta} = \boldsymbol{E}_{x}\left\{\left[\nabla_{\theta}\log u_{\theta}(\boldsymbol{x})\right]\left[\nabla_{\theta}\log u_{\theta}(\boldsymbol{x})\right]^{\mathrm{T}}\right\} \,. \tag{5.11}$$

由于 $\boldsymbol{F}_{\boldsymbol{\theta}}$ 是半正定的，所以存在楚列斯基（Cholesky）分解，即

$$\boldsymbol{F}_{\theta} = \boldsymbol{L}_{\theta}^{\mathrm{T}}\boldsymbol{L}_{\theta} \,, \tag{5.12}$$

由式（5.10）和式（5.12）可得

$$k_F(\boldsymbol{X},\boldsymbol{Y})=\left(\boldsymbol{g}_\theta^X\right)^{\mathrm{T}}\boldsymbol{g}_\theta^Y, \tag{5.13}$$

其中

$$\boldsymbol{g}_\theta^X=\boldsymbol{L}_\theta\boldsymbol{G}_\theta^X=\boldsymbol{L}_\theta\nabla_\theta\log u_\theta(\boldsymbol{X}),\quad \boldsymbol{g}_\theta^Y=\boldsymbol{L}_\theta\boldsymbol{G}_\theta^Y=\boldsymbol{L}_\theta\nabla_\theta\log u_\theta(\boldsymbol{Y}) \tag{5.14}$$

分别称为 $\boldsymbol{X}$ 和 $\boldsymbol{Y}$ 的费希尔向量，其维数与对应的梯度向量相同。

若概率密度函数 $u_\theta(\boldsymbol{X})$是由 GMM 方法得到，且其第 $i$ 个高斯成分为 $u_i(\boldsymbol{X})$，则对于任意样本特征（描述子）$\boldsymbol{x}_m$，其来自第 $i$ 个高斯成分的概率为

$$\pi_m(i)=\frac{w_iu_i(\boldsymbol{x}_m)}{\sum_{j=1}^{p}w_ju_j(\boldsymbol{x}_m)}, \tag{5.15}$$

其中，$p$ 为 GMM 的阶数；$w_i$ 为其中第 $i$ 个高斯成分的权重。

对 GMM 成分的权重、均值向量和标准差向量（对协方差矩阵为对角阵的情况）求偏导数，则有

$$\boldsymbol{g}_{w,i}^X=\frac{1}{N\sqrt{w_i}}\sum_{m=1}^{N}[\pi_m(i)-w_i], \tag{5.16}$$

$$\boldsymbol{g}_{\boldsymbol{\mu},i}^X=\frac{1}{N\sqrt{w_i}}\sum_{m=1}^{N}\pi_m(i)\left(\frac{\boldsymbol{x}_m-\boldsymbol{\mu}_i}{\boldsymbol{\sigma}_i}\right), \tag{5.17}$$

$$\boldsymbol{g}_{\boldsymbol{\sigma},i}^X=\frac{1}{N\sqrt{2w_i}}\sum_{m=1}^{N}\pi_m(i)\left[\frac{(\boldsymbol{x}_m-\boldsymbol{\mu}_i)^2}{\boldsymbol{\sigma}_i^2}-1\right]. \tag{5.18}$$

于是得到特征集 $\boldsymbol{X}$ 的费希尔向量编码

$$\boldsymbol{g}_\theta^X=(g_{w,i}^X,g_{\boldsymbol{\mu},i}^X,g_{\boldsymbol{\sigma},i}^X), \tag{5.19}$$

其维数为$(2M+1)p$，又由于各高斯成分权值之和为 1，所以最终费希尔向量编码的维数为$(2M+1)p-1$。

当 $N=1$ 时，仅对一个特征描述子进行费希尔向量编码，并将其从 $M$ 维映射到了$(2M+1)p-1$ 维；当 $N$ 是整个图像或视频中的所有局部特征描述子的个数时，其相当于将整个图像或视频编码为一个超向量，并满足输入鉴别型分类器中要求特征维数一致的条件。此外，其对原始特征向量的每一维都参与均值和方差的计算，充分包含了生成式建模过程中的结构性信息，对图像或视频的表达更加细致。

2. 基于 KNR 编码和 BoVW 模型的视频目标行为识别框架

对 DOFT 之轨迹特征描述子，或其他描述子如 HOG、HOF、MBH，可通过

式（5.7）进行 KNR 编码，得到对应的高维描述子（例如，设 $N$ 个原始密集轨迹的 HOG 描述子集合为 $\{\boldsymbol{x}_1, \boldsymbol{x}_2, \cdots, \boldsymbol{x}_N\} \in R^{M\times N}$，经过 k-means 聚类后可得到 $M_1$ 个聚类中心 $\{\boldsymbol{z}_i\}_{i=1}^{M_1}$，再经过 KNR 编码后得到 KNRc 向量集 $\{\boldsymbol{\gamma}_{\boldsymbol{x}_1}, \boldsymbol{\gamma}_{\boldsymbol{x}_2}, \cdots, \boldsymbol{\gamma}_{\boldsymbol{x}_N}\} \in R^{M_1\times N}$），然后借助于视觉词袋（bag of visual words，BoVW）模型[490, 502]形成特征向量，最后输入分类器（如线性 SVM）进行分类。于是得到如图 5.19 所示的视频目标行为识别框架[211, 499]。

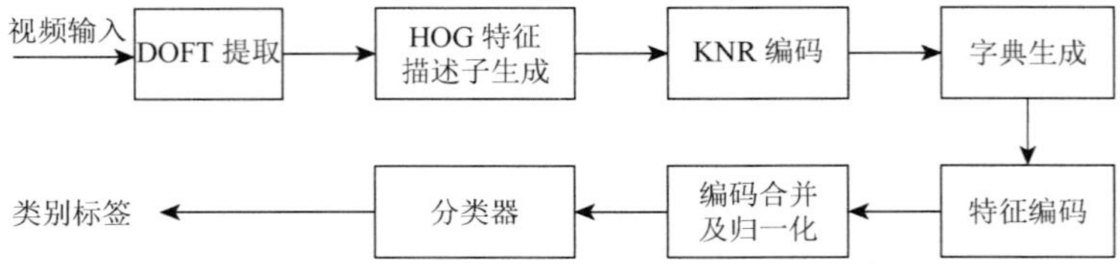


图 5.19　基于 KNR 编码和 BoVW 模型的视频目标行为识别框架[211, 499]

为验证该框架的行为识别有效性，这里利用魏茨曼（Weizmann）公开数据库[503]进行实验。该数据库中视频的拍摄背景和视角不变，摄像头静止。数据库包括 90 段视频，分别录制 9 个人的弯腰（bend）、摇摆（jack）、跳跃（jump）、奔跑（run）、屈体跳（pjump）、旁侧支撑（side）、溜走（skip）、散步（walk）、摇摆 1（wave1）和摇摆 2（wave2）等 10 个不同动作，其中典型数据如图 5.20[211, 503]所示。

(a) 旁侧支撑　(b) 跳跃　(c) 溜走　(d) 摇摆1

(e) 弯腰　(f) 摇摆2　(g) 屈体跳　(h) 奔跑

图 5.20　Weizmann 数据库中的部分典型行为

由于影响 KNR 编码结果的因素包括核函数及其参数、聚类中心个数等，其中核函数及其参数的选取决定了映射后的特征空间是否可以恰当表达原始数据特征，聚类中心个数 $M_1$ 决定了映射后的高维特征维数，$M_1$ 过大会增加计算的复杂度，过小则达不到提取更多鉴别信息的目的，所以这里重点考察这些因素的影响。

实验中取密集轨迹采样条数 $N = 20000$ 条，字典生成的“视觉词汇”数 $M_2 = 2000$ 个，无 PCA 特征预处理，编码合并使用求和形式，归一化中先进行指数归一化再进行二范数归一化。

首先，考察无 KNR 编码和不同核函数下 KNR 编码对识别率的影响。实验中分别选用四种核函数用于 KNR 编码：归一化线性核 $k(\boldsymbol{x}, \boldsymbol{z}) = \langle \boldsymbol{x}/\|\boldsymbol{x}\|, \boldsymbol{z}/\|\boldsymbol{z}\| \rangle$，多项式核 $k(\boldsymbol{x}, \boldsymbol{z}) = (\langle \boldsymbol{x}, \boldsymbol{z} \rangle + c)^d$，单位圆上的高斯核 $k(\boldsymbol{x}, \boldsymbol{z}) = \|\boldsymbol{x}\|\ \|\boldsymbol{z}\|\ \exp\{-\alpha(\boldsymbol{x}/\|\boldsymbol{x}\| - \boldsymbol{z}/\|\boldsymbol{z}\|)^2\}$，以及 arc-cosine 核：

$$k_{Id}(\boldsymbol{x}, \boldsymbol{z}) = \begin{cases} 1 - \dfrac{\varphi}{\pi}, & Id = 0, \\ \dfrac{\|\boldsymbol{x}\|\|\boldsymbol{z}\|}{\pi}[\sin\varphi + (\pi - \varphi)\cos\varphi], & Id = 1, \end{cases} \tag{5.20}$$

其中，$\varphi = \arccos(\langle \boldsymbol{x}, \boldsymbol{z} \rangle / \|\boldsymbol{x}\|\ \|\boldsymbol{z}\|)$。

取参数 $c = 1$、$d = 3$、$\alpha = 0.6$、$M_1 = 200$ 进行实验，实验结果如表 5.8 所示（其中 MBHx 和 MBHy 分别表示 $x$ 和 $y$ 方向的 MBH）[211, 499]。

**表 5.8 无 KNR 编码和不同核函数下 KNR 编码的行为识别率**

| 核函数种类 | 行为识别率/% | | | | |
|---|---|---|---|---|---|
| | HOG | HOF | MBHx | MBHy | 4-Fusion |
| 无 KNR 编码 | 71.94 | 65.28 | 67.22 | 59.44 | 68.33 |
| 归一化线性核 | 80 | **75.28** | 73.33 | 78.89 | **79.72** |
| 多项式核 | 65.28 | 65 | 63.33 | 64.72 | 68.33 |
| 单位圆上的高斯核 | 73.33 | 71.11 | 78.61 | **81.39** | 78.89 |
| arc-cosine 核（$Id = 0$） | **81.29** | 72.56 | **80.22** | 78.66 | 78.56 |
| arc-cosine 核（$Id = 1$） | 75.48 | 71.11 | 70.15 | 65.56 | 70.22 |

由表 5.8 可以看到：使用 KNR 编码可以显著提升行为识别率；在 HOG、HOF 和 4-Fusion 三种特征中，归一化线性核总体表现最佳，且该核函数无须核参数调节；在使用单位圆上的高斯核时，MBHy 特征相较于无 KNR 编码时行为识别率提升约 22%，而融合特征 4-Fusion 则提升了约 10.6%。

其次，以单位圆上的高斯核为例，考察核函数参数对行为识别率的影响。单位圆上高斯核相较于传统高斯核的优点在于核参数动态调节范围小，通常核参数 $\alpha$ 取在[0.1, 1]上取值。以间隔为 0.2 取不同的核参数值进行实验，结果如表 5.9[211, 499]所示。

**表 5.9　单位圆上的高斯核在不同核参数值下的行为识别率**

| 核参数 $\alpha$ | 行为识别率/% | | | | |
|---|---|---|---|---|---|
| | HOG | HOF | MBHx | MBHy | 4-Fusion |
| 0.2 | 66.94 | 61.94 | 67.78 | 60.00 | 63.61 |
| 0.4 | 71.11 | 66.74 | 70.65 | 70.56 | 68.78 |
| 0.6 | 73.33 | 71.11 | **78.61** | **81.39** | 78.89 |
| 0.8 | 76.67 | **75.83** | 74.72 | 78.89 | **80.00** |
| 1.0 | **76.94** | **75.83** | 68.61 | 75.83 | 75.83 |

由表 5.9 可以看到，每种特征对应的最优核参数不同：HOG 和 HOF 对应的核参数约为 1.0，MBHx 和 MBHy 在核参数约为 0.6 时达到最佳识别率，而融合特征 4-Fusion 的最优核参数约为 0.8。当核参数值小于 0.4 时，与表 5.8 中无 KNR 编码的识别率相比已无明显优势。

最后，以单位圆上的高斯核为例，考察 KNR 编码中聚类中心数对识别率的影响。使用单位圆上的高斯核函数（$\alpha = 0.6$）进行 KNR 编码，聚类中心数以间隔 100 从 100～500 取值，实验结果如表 5.10[211, 499]所示。

**表 5.10　单位圆上的高斯核（参数 $\alpha = 0.6$）在不同聚类中心数 $M_1$ 下的行为识别率**

| 聚类中心数 $M_1$ | 行为识别率/% | | | | |
|---|---|---|---|---|---|
| | HOG | HOF | MBHx | MBHy | 4-Fusion |
| 100 | 68.15 | 67.58 | 70.56 | 76.21 | 66.48 |
| 200 | 73.33 | 71.11 | **78.61** | **81.39** | 78.89 |
| 300 | 74.35 | 72.68 | 78.52 | 80.60 | **79.35** |
| 400 | 75.12 | **73.25** | 77.86 | 79.65 | 78.95 |
| 500 | **76.00** | 74.14 | 77.65 | 80.14 | 76.86 |

从表 5.10 可以看到，随着聚类中心数的增大，各特征对应的识别率在 $M_1 = 200$ 后无明显变化。所以，后续实验中将取 $M_1 = 200$。

总之，图 5.19 的识别框架较传统基于 BoVW 模型的行为识别性能有显著提升。

### 3. 基于 KNR 编码和费希尔向量编码的视频目标行为识别框架

费希尔向量编码和 KNR 编码都是打破了核函数的内聚性，将特征通过非线性映射到高维空间中并显式地表达出来。前者的优势在于可以将任意个数的特征描述子进行编码输出，而后者通常仅对单个特征描述子编码。据此可融合二者优势，构建如图 5.21 所示的基于 KNR 编码和费希尔向量编码的视频行为识别框架[211, 499]：特征描述子生成之后，送入 KNR 对单条视频轨迹特征进行编码，每个视频输出的 KNRc 向量集作为一个样本，经 GMM 建模后进行费希尔向量编码，完成视频目标行为识别[211, 499]。

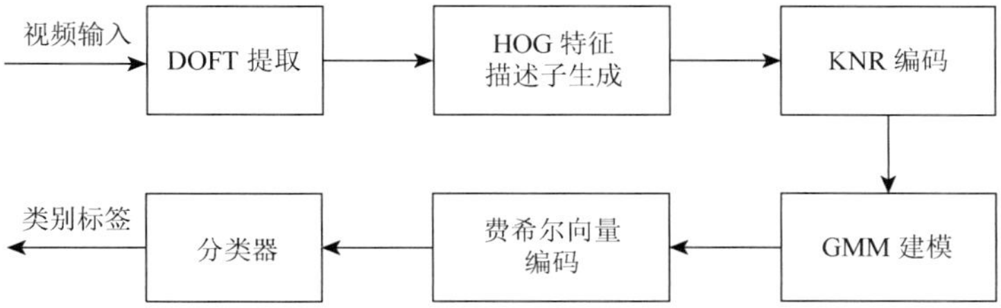


图 5.21 基于 KNR 编码和费希尔向量编码的视频目标行为识别框架[211, 499]

同样利用魏茨曼数据库进行实验，验证图 5.21 之框架的行为识别有效性。其中，重点探究 GMM 阶数 $p$ 对识别率的影响，以及不同 KNR 核函数对识别率的影响。

首先，考察 GMM 建模中模型规模（高斯成分个数）$p$ 对识别率的影响。此实验不设置 KNR 编码层，直接对原始特征描述子进行 GMM 建模和费希尔向量编码，GMM 中阶数 $p$ 设为 16、32、64、128 和 256，实验结果如表 5.11 所示[211, 499]。

**表 5.11 不同模型阶数 $p$ 下 GMM 的行为识别率**

| $p$ | 行为识别率/% | | | | |
|---|---|---|---|---|---|
| | HOG | HOF | MBHx | MBHy | 4-Fusion |
| 16 | 74.44 | 74.44 | 74.17 | 78.61 | 77.50 |
| 32 | 77.50 | 77.50 | 74.44 | 78.33 | **80.83** |
| 64 | 80.00 | 79.72 | 76.67 | 80.83 | 78.89 |
| 128 | **81.11** | 80.00 | 76.67 | **81.89** | 80.00 |
| 256 | 80.00 | **82.22** | **76.67** | 81.67 | 78.89 |

从表 5.11 可以看到，随着 GMM 规模的增加，各特征的行为识别率也呈上升趋势，且在 $p = 128$ 时基于 HOG、MBHy 和融合特征 4-Fusion 呈饱和趋势，而基于 HOF 和 MBHx 特征的识别率在 $p = 256$ 时最佳。为此，将 $p = 256$ 作为后续实验的基础参数。

另外，考察不同核函数下 KNR 编码对行为识别率的影响。实验中设置 GMM 阶数为 $p = 256$，分别在无 KNR 编码和使用上述 4 个核函数（其中高斯核参数 $\alpha = 0.6$，多项式核参数 $c = 1$、$d = 3$）进行 KNR 编码，实验结果如表 5.12 所示[211, 499]。

**表 5.12　无 KNR 编码和不同核函数下 KNR 编码对基于费希尔向量编码的行为识别率**

| 核函数种类 | 行为识别率/% | | | | |
|---|---|---|---|---|---|
| | HOG | HOF | MBHx | MBHy | 4-Fusion |
| 无 KNR 编码 | 80.00 | 82.22 | 76.67 | 81.67 | 78.89 |
| 归一化线性核 | 80.83 | 80.83 | 80.33 | **84.17** | **83.06** |
| 多项式核 | 77.62 | 75.84 | 70.52 | 76.88 | 77.85 |
| 单位圆上的高斯核 | 56.11 | 55.28 | 58.78 | 53.06 | 66.19 |
| arc-cosine 核（$Id = 0$） | **83.06** | **83.06** | **80.83** | **84.17** | 81.94 |
| arc-cosine 核（$Id = 1$） | 58.06 | 56.11 | 55.00 | 56.39 | 58.89 |

从表 5.12 可以看到，除了高斯核、多项式核以及 arc-cosine 核在 $Id = 1$ 时表现不佳以外，归一化线性核与 arc-cosine 核在 $Id = 0$ 时较于传统（无 KNR）费希尔向量编码的行为识别率总体上有较大提升。

## 5.4　本 章 小 结

本章以基于 KNR 直方图拟合与分解的图像分割、基于 KNR 曲面拟合与再采样的图像放大、基于 KNR 坐标映射配准和 KPPLR 曲面拟合的多帧融合图像超分辨重建、基于 KNR 波形估计和 KND 端点检测及 KPPLR 重构的语音增强为实例，以及以 KNR、KND、KNRD 为基础的手写数字识别、人脸识别、说话人识别、雷达目标识别和视频目标行为识别为案例，介绍了本书有关理论和算法在信号分析处理中的应用和模式识别中的应用，为更多的场景应用提供了典型参照。

# 参 考 文 献


[1] McCulloch W S, Pitts W. A logical calculus of the ideas immanent in nervous activity[J]. The Bulletin of Mathematical Biophysics, 1943, 5 (4): 115-133.

[2] Hebb D O. The Organization of Behavior: A Neuropsychological Theory[M]. New York: Wiley, 1949.

[3] Minsky M L. Theory of Neural-analog Reinforcement Systems and Its Application to the Brain Model Problem[M]. New Jersey: Princeton University Press, 1954.

[4] Von Neumann J. The General and Logical Theory of Automata[M]//Jeffress L A. Cerebral Mechanisms in Behavior-the Hixon Symposium. New York: John Wiley & Sons, 1951.

[5] Turing A M. Computing machinery and intelligence[J]. Mind, 1950, 49: 433-460.

[6] Minsky M. A selected descriptor-indexed bibliography to the literature on artificial intelligence[J]. IRE Transactions on Human Factors in Electronics, 1961, HFE-2 (1): 39-55.

[7] Feigenbaum E. Artificial intelligence research[J]. IEEE Transactions on Information Theory, 1963, 9 (4): 248-253.

[8] McCarthy J, Minsky M, Rochester N, et al. A proposal for the Dartmouth summer research project on artificial intelligence, August 31, 1955[J]. AI Magazine, 2006, 27 (4): 12-14.

[9] Andresen S L. John McCarthy: Father of AI[J]. IEEE Intelligent Systems, 2002, 17 (5): 84-85.

[10] Solomonoff R J. Some recent work in artificial intelligence[J]. Proceedings of the IEEE, 1966, 54 (12): 1687-1697.

[11] Marr D. Artificial intelligence: A personal view[J]. Artificial Intelligence, 1977, 9 (1): 37-48.

[12] Nilsson N J. Principles of Artificial Intelligence[M]. Palo Alto, California: Tioga Press, 1980.

[13] Simmons A B, Chappell S G. Artificial intelligence: Definition and practice[J]. IEEE Journal of Oceanic Engineering, 1988, 13 (2): 14-42.

[14] Simon H A. Artificial intelligence: Where has it been, and where is it going?[J]. IEEE Transactions on Knowledge and Data Engineering, 1991, 3 (2): 128-136.

[15] Abbass H. Editorial: What is artificial intelligence?[J]. IEEE Transactions on Artificial Intelligence, 2021, 2 (2): 94-95.

[16] Copeland B J. Artificial intelligence.[J/OL]. Encyclopedia Britannica. (2022-10-10) [2024-05- 01]. https://www.britannica. com/technology/artificial-intelligence. Accessed on October 10, 2022.

[17] Newell A, Simon H. The logic theory machine: A complex information processing system[J]. IRE Transactions on Information Theory, 1956, 2 (3): 61-79.

[18] Minsky M. Heuristic Aspects of the Artificial Intelligence Problem[R]. Cambridge: MIT Lincoln Laboratory, 1956.

[19] Minsky M. Steps toward artificial intelligence[J]. Proceedings of the IRE, 1961, 49 (1): 8-30.

[20] Simon H A, Newell A. Heuristic problem solving: The next advance in operations research[J]. Operations Research, 1958, 6 (1): 1-10.

[21] Gelernter H L, Rochester N. Intelligent behavior in problem-solving machines[J]. IBM Journal of Research and Development, 1958, 2 (4): 336-345.

[22] Gelernter H L. Realization of a geometry theorem proving machine[C]//Proceedings of the 1st International Conference on Information Processing, Paris, France, 1959: 273-282.

[23] Newell A, Shaw J C, Simon H A. Report on general problem-solving program[C]//Proceedings of the 1st International Conference on Information Processing, Paris, France, 1959: 256-264.

[24] Slagle J R. A heuristic program that solves symbolic integration problems in freshman calculus[J]. Journal of the ACM, 1963, 10 (4): 507-520.

[25] Shannon C E. Programming a computer for playing chess[J]. The London, Edinburgh, and Dublin Philosophical Magazine and Journal of Science, 1950, 41 (314): 256-275.

[26] Strachey C S. Logical or non-mathematical programmes[C]//Proceedings of the 1952 ACM National Meeting (Toronto). New York: ACM, 1952: 46-49.

[27] Newell A, Shaw J C, Simon H A. Chess-playing programs and the problem of complexity[J]. IBM Journal of Research and Development, 1958, 2 (4): 320-335.

[28] Samuel A L. Some studies in machine learning using the game of checkers[J]. IBM Journal of Research and Development, 1959, 3 (3): 210-229.

[29] Samuel A L. Some studies in machine learning using the game of checkers. II: Recent progress[J]. IBM Journal of Research and Development, 1967, 11 (6): 601-617.

[30] Wall R E. Some of the engineering aspects of the machine translation of language[J]. Transactions of the American Institute of Electrical Engineers, Part I: Communication and Electronics, 1956, 75 (5): 580-585.

[31] Bonney R B. A universal computer language translator[C]//Proceedings of the May 6-8, 1958, Western Joint Computer Conference on Contrasts in Computer. New York: ACM, 1958: 230-233.

[32] Kosarin M G. Language translation by machine[J]. Journal of the SMPTE, 1959, 68 (4): 232-233.

[33] Weis Jr E B, Lajeunesse D J. SYSTRAN (system analysis translator): A digital computer program[R]. The 6570th Technical Report of Aerospace Medical Research Laboratories (Supplement I), Number AMRL-TR-65-133, 1965: 1-83.

[34] Green B F Jr, Wolf A K, Chomsky C, et al. Baseball: An automatic question-answerer [C]//Papers Presented at The May 9-11, 1961, Western Joint IRE-AIEE-ACM Computer Conference. New York: ACM, 1961: 219-224.

[35] Simmons R F. Answering English questions by computer: A survey[J]. Communications of the ACM, 1965, 8 (1): 53-70.

[36] Lindsay R K. Inferential memory as the basis of machines which understand natural language [M]//Feigenbaum E A, Feldman J. Computers and Thought. New York: McGraw Hill Book Company, Inc., 1963: 217-233.

[37] Weizenbaum J. ELIZA: A computer program for the study of natural language communication

between man and machine[J]. Communications of the ACM, 1966, 9 (1): 36-45.

[38] Feigenbaum E A. The simulation of verbal learning behavior[C]//Papers presented at the May 9-11, 1961, Western Joint IRE-AIEE-ACM Computer Conference. New York: ACM, 1961: 121-132.

[39] Feldman J. Simulation of behavior in the binary choice experiment[C]//Papers presented at the May 9-11, 1961, Western Joint IRE-AIEE-ACM Computer Conference. New York: ACM, 1961: 133-144.

[40] Selfridge O G. Pattern recognition and modern computers[C]//Proceedings of the March 1-3, 1955, Western Joint Computer Conference. New York: ACM, 1955: 91-93.

[41] Dinneen G P. Programming pattern recognition[C]//Proceedings of the March 1-3, 1955, Western Joint Computer Conference. New York: ACM, 1955: 94-100.

[42] Rosenblatt F. The perceptron: A probabilistic model for information storage and organization in the brain[J]. Psychological Review, 1958, 65 (6): 386-408.

[43] Unger S H. Pattern detection and recognition[J]. Proceedings of the IRE, 1959, 47 (10): 1737-1752.

[44] Frank R. Principles of Neurodynamics: Perceptrons and the Theory of Brain Mechanisms[M]. Washington: Spartan Books, 1961.

[45] Selfridge O G , Neisser U. Pattern recognition by machine[J]. Scientific American, 1960, 203 (2): 60-68.

[46] Keller H B. Finite automata, pattern recognition and perceptrons[J]. Journal of the ACM, 1961, 8 (1): 1-20.

[47] Uhr L. "Pattern recognition" computers as models for form perception[J]. Psychological Bulletin, 1963, 60: 40-73.

[48] Von Foerster H. Computation in neural nets[J]. Biosystems, 1967, 1 (1): 47-93.

[49] Taylor W K. Machine learning and recognition of faces[J]. Electronics Letters, 1967, 3 (9): 436-437.

[50] Amari S. A theory of adaptive pattern classifiers[J]. IEEE Transactions on Electronic Computers, 1967, EC-16 (3): 299-307.

[51] Fukushima K. Visual feature extraction by a multilayered network of analog threshold elements[J]. IEEE Transactions on Systems Science and Cybernetics, 1969, 5 (4): 322-333.

[52] Ziff P. The feelings of robots[J]. Analysis, 1959, 19(3): 64-68.

[53] Licklider J C R. Man-computer symbiosis[J]. IRE Transactions on Human Factors in Electronics, 1960, 1 (1): 4-11.

[54] David E E, Selfridge O G. Eyes and ears for computers[J]. Proceedings of the IRE, 1962, 50 (5): 1093-1101.

[55] Ernst H A. MH-1, a computer-operated mechanical hand[C]//Proceedings of the May 1-3, 1962, Spring Joint Computer Conference. New York: ACM, 1962: 39-51.

[56] Putman H. Robots: machines or artificially created life? [J]. The Journal of Philosophy, 1964, 61 (21): 668.

[57] Clowes M B. On seeing things[J]. Artificial Intelligence, 1971, 2 (1): 79-116.

[58] McCarthy J, Earnest L D, Reddy D R, et al. A computer with hands, eyes, and

ears[C]//Proceedings of the December 9-11, 1968, Fall Joint Computer Conference, Part I. New York: ACM, 1968: 329-338.
[59] McCarthy J. Programs with common sense[C]//Proceedings of the Symposium on Mechanization of Thought Processes. London, National physical Laboratory，HMSO，1958, volume 1: 77-91.
[60] McCarthy J. LISP: A programming system for symbolic manipulations[C]//Preprints of Papers Presented at the 14th National Meeting of the Association for Computing Machinery. New York: ACM, 1959: 1-4.
[61] McCarthy J. History of LISP[M]//Wexelblat R L. History of Programming Languages. Academic Press, 1978: 173-185.
[62] Davis R E. Logic programming and prolog: A tutorial[J]. IEEE Software, 1985, 2 (5): 53-62.
[63] Wang H. A variant to turing's theory of computing machines[J]. Journal of the ACM, 1957, 4 (1): 63-92.
[64] Burks A W, Wang H. The logic of automata: Part I[J]. Journal of the ACM, 1957, 4 (2): 193-238.
[65] Burks A W, Wang H. The logic of automata: Part II[J]. Journal of the ACM, 1957, 4 (3): 279-297.
[66] Wang H. Proving theorems by pattern recognition I[J]. Communications of the ACM, 1960, 3 (4): 220-234.
[67] Lederberg J, Feigenbaum E A. Mechanization of Inductive Inference in Organic Chemistry[M]// Kleinmuntz B. Formal Representation of Human Judgment. New York: Wiley, 1968: 187-267.
[68] Rosen C A, Nilsson N. Application of intelligent automata to reconnaissance[R]. Menlo Park: Stanford Research Institute, 1968.
[69] Lederberg J. Dendral-64: A system for computer construction, enumeration and notation of organic molecules as tree structures and cyclic graphs, Part Ⅱ. Topology of cyclic graphs[R]. Washington: NASA, 1965.
[70] Nilsson N. A mobile automaton: An application of artificial intelligence techniques[C]// Proceedings of the 1st International Joint Conference on Artificial Intelligence. Washington, 1969: 509-520.
[71] Raphael B. Robot research at Stanford research institute[R]. Menlo Park: Stanford Research Institute, 1972.
[72] Kuipers B, Feigenbaum E A, Hart P E, et al. Shakey: From conception to history[J]. AI Magazine, 2017, 38 (1): 88-103.
[73] Bell A, Quillian M R. Capturing concepts in a semantic net[R]. Massachusetts: Air Force Cambridge Research Laboratory, 1969.
[74] Fikes R E, Nilsson N J. STRIPS: A new approach to the application of theorem proving to problem solving[J]. Artificial Intelligence, 1971, 2 (3/4): 189-208.
[75] Winograd T A. Procedures as a representation for data in a computer program for understanding natural language[D]. Cambridge: Massachusetts Institute of Technology, 1971.
[76] Winograd T. Understanding natural language[J]. Cognitive Psychology, 1972, 3 (1): 1-191.

[77] Colmerauer A, Roussel P. The birth of Prolog[J]. SIGPLAN Notices of the ACM, 1993, 28 (3): 37-52.

[78] Kato I, Ohteru S, Kobayashi H, et al. Information-Power Machine with Senses and Limbs[M]// On Theory and Practice of Robots and Manipulators. Vienna: Springer, 1974: 11-24.

[79] Newell A, Simon H A. Computer science as empirical inquiry: Symbols and search[J]. Communications of the ACM, 1976, 19 (3): 113-126.

[80] Simmons R F, Bruce B C. Some relations between predicate calculus and semantic net representations of discourse[C]//Proceedings of the 2nd International Joint Conference on Artificial Intelligence. New York: ACM, 1971: 524-530.

[81] Winston P H. Learning structural descriptions from examples[R]. Cambridge: Department of Electrical Engineering and Computer Science, MIT, 1970.

[82] Huffman D A. Impossible Objects as Nonsense Sentences[M]//Meltzer B, Michie D. Machine Intelligence 6. Edinburgh: Edinburgh University Press, 1971: 295-324.

[83] Nakano K. Associatron: a model of associative memory[J]. IEEE Transactions on Systems, Man, and Cybernetics, 1972, SMC-2 (3): 380-388.

[84] Amari S I. Characteristics of random nets of analog neuron-like elements[J]. IEEE Transactions on Systems, Man, and Cybernetics, 1972, SMC-2 (5): 643-657.

[85] Kohonen T. Correlation matrix memories[J]. IEEE Transactions on Computers, 1972, C-21 (4): 353-359.

[86] Kohonen T. An adaptive associative memory principle[J]. IEEE Transactions on Computers, 1974, C-23 (4): 444-445.

[87] Fukushima K. Cognitron: A self-organizing multilayered neural network[J]. Biological Cybernetics, 1975, 20 (3/4): 121-136.

[88] Albus J S, Evans J M. Robot systems[J]. Scientific American, 1976, 234 (2): 76-86.

[89] Amari S I. Neural theory of association and concept-formation[J]. Biological Cybernetics, 1977, 26 (3): 175-185.

[90] Fukushima K. A model of associative memory in the brain[J]. Kybernetik, 1973, 12 (2): 58-63.

[91] Fukushima K, Miyake S. A self-organizing neural network with a function of associative memory: Feedback-type cognitron[J]. Biological Cybernetics, 1978, 28 (4): 201-208.

[92] Kohonen T, Reuhkala E, Mäkisara K, et al. Associative recall of images[J]. Biological Cybernetics, 1976, 22 (3): 159-168.

[93] Kohonen T, Oja E. Fast adaptive formation of orthogonalizing filters and associative memory in recurrent networks of neuron-like elements[J]. Biological Cybernetics, 1976, 21 (2): 85-95.

[94] Felgenbaum E A. The art of artificial intelligence: themes and case studies of knowledge engineering[C]//Proceedings of the 5th International Joint Conference on Artificial Intelligence. New York: ACM, 1977: 1014-1029.

[95] Reddy D R, Erman L D, Fennell R D, et al. The hearsay-I speech understanding system: an example of the recognition process[J]. IEEE Transactions on Computers, 1976, C-25 (4): 422-431.

[96] Erman L D, Hayes-Roth F, Lesser V R, et al. The hearsay-Ⅱ speech-understanding system: integrating knowledge to resolve uncertainty[J]. ACM Computing Surveys, 1980, 12 (2):

213-253.
[97] Shortliffe E H, Davis R, Axline S G, et al. Computer-based consultations in clinical therapeutics: Explanation and rule acquisition capabilities of the MYCIN system[J]. Computers and Biomedical Research, 1975, 8 (4): 303-320.
[98] Appel K, Haken W. The solution of the four-color-map problem[J]. Scientific American, 1977, 237 (4): 108-121.
[99] Minsky M, Papert S . Perceptrons: An Introduction to Computational Geometry[M]. Cambridge: MIT Press, 1969.
[100] Hopfield J J. Neural networks and physical systems with emergent collective computational abilities[J]. Proceedings of the National Academy of Sciences of the United States of America, 1982, 79 (8): 2554-2558.
[101] Werbos P J. Beyond Regression: New Tools for Prediction and Analysis in the Behavioral Sciences[D]. Cambridge: PhD Thesis of Harvard University, 1974.
[102] LeCun Y. A learning scheme for asymmetric threshold networks[J]. Proceedings of Cognitiva, 1985: 599-604.
[103] Ackley D, Hinton G , Sejnowski T. A learning algorithm for boltzmann machines[J]. Cognitive Science, 1985, 9 (1): 147-169.
[104] Rumelhart D E, Hinton G E, Williams R J. Learning representations by back-propagating errors[J]. Nature, 1986, 323 (6088): 533-536.
[105] LeCun Y, Boser B, Denker J S, et al. Backpropagation applied to handwritten zip code recognition[J]. Neural Computation, 1989, 1 (4): 541-551.
[106] Feigenbaum E A. Knowledge engineering: The applied side of artificial intelligence[J]. Annals of the New York Academy of Sciences, 1984, 426: 91-107.
[107] Lenat D, Prakash M, Shepherd M. CYC: Using common sense knowledge to overcome brittleness and knowledge acquistion bottlenecks[J]. AI Magazine, 1985, 6 (4): 65-85.
[108] Chandrasekaran B. Generic tasks in knowledge-based reasoning: high-level building blocks for expert system design[J]. IEEE Expert, 1986, 1 (3): 23-30.
[109] Feigenbaum E, Buchanan B. Heuristic programming project: October 1979-September 1982[R]. Stamford: Knowledge Systems Laboratory, Stanford University, 1985.
[110] Kraft A. XCON: An Expert Configuration System at Digital Equipment Corporation[M]// Winston P H, Prendergast K A.The AI Business. Cambridge: The MIT Press, 1984: 41-50.
[111] Fox M S, Smith S F. ISIS: A knowledge-based system for factory scheduling[J]. Expert Systems, 1984, 1 (1): 25-49.
[112] Hollan J D, Hutchins E L, Weitzman L. STEAMER: An interactive inspectable simulation-based training system[J]. AI Magazine, 1984, 5 (2): 15-27.
[113] Sviokla J J. Business implications of knowledge-based systems[J]. ACM SIGMIS Database: the DATABASE for Advances in Information Systems, 1986, 17 (4): 5-19.
[114] 张钹. 近十年人工智能的进展[J]. 模式识别与人工智能, 1995, 8 (S1): 1-9.
[115] Brooks R A. Achieving artificial intelligence through building robots[R]. Cambridge: MIT, 1986.

[116] Brooks R. A robust layered control system for a mobile robot[J]. IEEE Journal on Robotics and Automation, 1986, 2 (1): 14-23.
[117] Brooks R A. A robot that walks: Emergent behaviors from a carefully evolved network[J]. Neural Computation, 1989, 1 (2): 253-262.
[118] Brooks R A. Elephants don't play chess[J]. Robotics and Autonomous Systems, 1990, 6 (1/2): 3-15.
[119] Selfridge O G, Franklin J A. The perceiving robot: What does it see? What does it do? [C]// Proceedings of 5th IEEE International Symposium on Intelligent Control. New York: IEEE, 1990: 146-151.
[120] Suchman L A. Plans and Situated Actions: The Problem of Human-machine Communication[M]. Cambridge: Cambridge University Press, 1987.
[121] Waltz M, Fu K. A heuristic approach to reinforcement learning control systems[J]. IEEE Transactions on Automatic Control, 1965, 10 (4): 390-398.
[122] Fu K S. Learning control systems: Review and outlook[J]. IEEE Transactions on Automatic Control, 1970, 15 (2): 210-221.
[123] Barto A G, Sutton R S, Brouwer P S. Associative search network: a reinforcement learning associative memory[J]. Biological Cybernetics, 1981, 40 (3): 201-211.
[124] Williams R J. Simple statistical gradient-following algorithms for connectionist reinforcement learning[J]. Machine Learning, 1992, 8 (3): 229-256.
[125] Vapnik V N. The Nature of Statistical Learning Theory[M]. New York: Springer New York, 1995.
[126] Kohonen T. The self-organizing map[J]. Proceedings of the IEEE 1990, 78 (9): 1464-1480.
[127] Specht D F. A general regression neural network[J]. IEEE Transactions on Neural Networks, 1991, 2 (6): 568-576.
[128] Atkeson C G, Moore A W, Schaal S. Locally weighted learning[J]. Artificial Intelligence, 1997, 11 (1): 11-73.
[129] Hinton G E. How neural networks learn from experience[J]. Scientific American, 1992, 267 (3): 144-151.
[130] Cortes C, Vapnik V N. Support-vector networks[J]. Machine Learning, 1995, 20 (3): 273-297.
[131] Pal S K, Mitra S. Multilayer perceptron, fuzzy sets, and classification[J]. IEEE Transactions on Neural Networks, 1992, 3 (5): 683-697.
[132] Yao X. Evolving artificial neural networks[J]. Proceedings of the IEEE, 1999, 87 (9): 1423-1447.
[133] Lave J, Wenger E. Situated Learning: Legitimate Peripheral Participation[M]. Cambridge: Cambridge University Press, 1991.
[134] Schlimmer J C, Hermens L A. Software agents: Completing patterns and constructing user interfaces[J]. Journal of Artificial Intelligence Research, 1993, 1: 61-89.
[135] Wooldridge M, Jennings N R. Intelligent agents: Theory and practice[J]. The Knowledge Engineering Review, 1995, 10 (2): 115-152.
[136] Wallace R. Artificial linguistic internet computer entity (Alice)[EB/OL]. [2022-12-31].

https://www.chatbots.org/chatbot/a.l.i.c.e/.
[137] McDermott D. How intelligent is Deep Blue?[N]. New York Times, 1997-05-14.
[138] Decoste D. The future of chess-playing technologies and the significance of Kasparov versus Deep Blue[R]. New York: AAAI, 1997.
[139] Lund H H, Bjerre C, Nielsen J H, et al. Adaptive LEGO robots. A robot = human view on robotics[C]//IEEE SMC'99 Conference Proceedings. 1999 IEEE International Conference on Systems, Man, and Cybernetics. New York: IEEE, 1999: 1017-1023.
[140] Dautenhahn K. The Lemur's tale-story-telling in Primates and other socially intelligent agents[R]. New York: AAAI, 1999.
[141] Sakagami Y, Watanabe R, Aoyama C, et al. The intelligent ASIMO: system overview and integration[C]//IEEE/RSJ International Conference on Intelligent Robots and Systems. New York: IEEE, 2002: 2478-2483.
[142] Glick J. AI in the news[J]. AI Magazine, 2002, 23 (1): 128.
[143] Thrun S, Montemerlo M, Dahlkamp H, et al. Stanley: The robot that won the DARPA Grand Challenge[J]. Journal of Field Robotics, 2006, 23 (9): 661-692.
[144] Petrovskaya A, Thrun S. Model based vehicle detection and tracking for autonomous urban driving[J]. Autonomous Robots, 2009, 26 (2): 123-139.
[145] Thrun S. Toward robotic cars[J]. Communications of the ACM, 2010, 53 (4): 99-106.
[146] Levinson J, Askeland J, Becker J, et al. Towards fully autonomous driving: Systems and algorithms[C]//2011 IEEE Intelligent Vehicles Symposium (IV). New York: IEEE, 2011: 163-168.
[147] Aron J. How innovative is Apple's new voice assistant, Siri？[J]. New Scientist, 2011, 212 (2836): 24, 29.
[148] Ferrucci D, Brown E, Chu-Carroll J, et al. Building Watson: an overview of the DeepQA project[J]. AI Magazine, 2010, 31 (3): 59-79.
[149] Ferrucci D A. Introduction to "this is Watson" [J]. IBM Journal of Research and Development, 2012, 56 (3/4): 1-15.
[150] Hinton G E, Salakhutdinov R R. Reducing the dimensionality of data with neural networks[J]. Science, 2006, 313 (5786): 504-507.
[151] Hinton G E, Osindero S, Teh Y W. A fast learning algorithm for deep belief nets[J]. Neural Computation, 2006, 18 (7): 1527-1554.
[152] Bengio Y, Lamblin P, Popovici D, et al. Greedy layer-wise training of deep networks [C]// Proceedings of the 19th International Conference on Neural Information Processing Systems. New York: ACM, 2006: 153-160.
[153] Hinton G E. Learning multiple layers of representation[J]. Trends in Cognitive Sciences, 2007, 11 (10): 428-434.
[154] Jain V, Seung H S. Natural image denoising with convolutional networks[C]//Proceedings of the 21st International Conference on Neural Information Processing Systems. New York: ACM, 2008: 769-776.
[155] Vincent P, Larochelle H, Bengio Y, et al. Extracting and composing robust features with

denoising autoencoders[C]//Proceedings of the 25th International Conference on Machine Learning. New York: ACM, 2008: 1096-1103.

[156] Raina R, Madhavan A, Ng A Y. Large-scale deep unsupervised learning using graphics processors[C]//Proceedings of the 26th Annual International Conference on Machine Learning. New York: ACM, 2009: 873-880.

[157] Lee H, Ekanadham C, Ng A Y. Sparse deep belief net model for visual area V2[C]//Proceedings of the 20th International Conference on Neural Information Processing Systems. New York: ACM, 2007: 873-880.

[158] Lee H, Grosse R, Ranganath R, et al. Convolutional deep belief networks for scalable unsupervised learning of hierarchical representations[C]//Proceedings of the 26th Annual International Conference on Machine Learning. New York: ACM, 2009: 609-616.

[159] Bengio Y. Learning Deep Architectures for AI[M]. Hanover, MA, USA: Now Publishers Inc, 2009.

[160] Le Roux N, Bengio Y. Representational power of restricted boltzmann machines and deep belief networks[J]. Neural Computation, 2008, 20 (6): 1631-1649.

[161] Vouk M A. Cloud computing: Issues, research and implementations[J]. Journal of Computing and Information Technology, 2008, 16 (4): 235-246.

[162] Wang L Z, Tao J, Kunze M, et al. Scientific cloud computing: Early definition and experience[C]//2008 10th IEEE International Conference on High Performance Computing and Communications. New York: IEEE, 2008: 825-830.

[163] Dikaiakos M D, Katsaros D, Mehra P, et al. Cloud computing: distributed Internet computing for IT and scientific research[J]. IEEE Internet Computing, 2009, 13 (5): 10-13.

[164] Armbrust M, Fox A, Griffith R, et al. A view of cloud computing[J]. Communications of the ACM, 2010, 53 (4): 50-58.

[165] Zhang Q, Cheng L, Boutaba R. Cloud computing: State-of-the-art and research challenges[J]. Journal of Internet Services and Applications, 2010, 1 (1): 7-18.

[166] Mell P, Grance T. The NIST definition of cloud computing[S]. Gaithersburg: National Institute of Standards and Technology, 2011: 1-7.

[167] Torralba A, Fergus R, Freeman W T. 80 million tiny images: A large data set for nonparametric object and scene recognition[J]. IEEE Transactions on Pattern Analysis and Machine Intelligence, 2008, 30 (11): 1958-1970.

[168] Li F F, Deng J, Li K. ImageNet: Constructing a large-scale image database[J]. Journal of Vision, 2010, 9 (8): 1037.

[169] Deng J, Dong W, Socher R, et al. ImageNet: A large-scale hierarchical image database[C]//2009 IEEE Conference on Computer Vision and Pattern Recognition. New York: IEEE, 2009: 248-255.

[170] Everingham M, Van Gool L, Williams C K I, et al. The PASCAL visual object classes (VOC)challenge[J]. International Journal of Computer Vision, 2010, 88 (2): 303-338.

[171] Xiao J X, Hays J, Ehinger K A, et al. SUN database: Large-scale scene recognition from abbey to zoo[C]//2010 IEEE Computer Society Conference on Computer Vision and Pattern

Recognition. New York: IEEE, 2010: 3485-3492.
[172] Schmidhuber J. Deep learning in neural networks: An overview[J]. Neural Networks, 2015, 61: 85-117.
[173] LeCun Y, Bengio Y, Hinton G. Deep learning[J]. Nature, 2015, 521 (7553): 436-444.
[174] Dean J, Corrado G, Monga R, et al. Large scale distributed deep networks[J]. Advances in Neural Information Processing Systems, 2012, 25 (2): 1223-1231.
[175] Krizhevsky A, Sutskever I, Hinton G E. ImageNet classification with deep convolutional neural networks[J]. Communications of the ACM, 2017, 60 (6): 84-90.
[176] Le Q V. Building high-level features using large scale unsupervised learning[C]//2013 IEEE International Conference on Acoustics, Speech and Signal Processing. New York: IEEE, 2013: 8595-8598.
[177] Russakovsky O, Deng J, Su H, et al. ImageNet: Large scale visual recognition challenge[J]. International Journal of Computer Vision, 2015, 115 (3): 211-252.
[178] Krizhevsky A, Sutskever I, Hinton G E. ImageNet classification with deep convolutional neural networks[J]. Communications of the ACM, 2017, 60 (6): 84-90.
[179] Purington A, Taft J G, Sannon S, et al. "Alexa is my new BFF" : Social roles, user satisfaction, and personification of the Amazon echo[C]//Proceedings of the 2017 CHI Conference Extended Abstracts on Human Factors in Computing Systems. New York: ACM, 2017: 2853-2859.
[180] Hoy M B. Alexa, Siri, Cortana, and more: An introduction to voice assistants[J]. Medical Reference Services Quarterly, 2018, 37 (1): 81-88.
[181] Lopatovska I, Rink K, Knight I, et al. Talk to me: Exploring user interactions with the Amazon Alexa[J]. Journal of Librarianship and Information Science, 2019, 51 (4): 984-997.
[182] Anthes E. Alexa, do I have COVID-19?[J]. Nature, 2020, 586 (7827): 22-25.
[183] Wang F Y, Zhang J J, Zheng X H, et al. Where does AlphaGo go: From church-turing thesis to AlphaGo thesis and beyond[J]. IEEE/CAA Journal of Automatica Sinica, 2016, 3 (2): 113-120.
[184] Silver D, Schrittwieser J, Simonyan K, et al. Mastering the game of Go without human knowledge[J]. Nature, 2017, 550 (7676): 354-359.
[185] Silver D, Hubert T, Schrittwieser J, et al. A general reinforcement learning algorithm that Masters chess, shogi, and Go through self-play[J]. Science, 2018, 362 (6419): 1140-1144.
[186] Schrittwieser J, Antonoglou I, Hubert T, et al. Mastering Atari, Go, chess and shogi by planning with a learned model[J]. Nature, 2020, 588 (7839): 604-609.
[187] AlQuraishi M. AlphaFold at CASP13[J]. Bioinformatics, 2019, 35 (22): 4862-4865.
[188] Shimizu H, Nakayama K I. Artificial intelligence in oncology[J]. Cancer Science, 2020, 111 (5): 1452-1460.
[189] Jumper J, Evans R, Pritzel A, et al. Highly accurate protein structure prediction with AlphaFold[J]. Nature, 2021, 596 (7873): 583-589.
[190] Roney J P, Ovchinnikov S. State-of-the-art estimation of protein model accuracy using AlphaFold[J]. Physical Review Letters, 2022, 129 (23): 238101.
[191] Nussinov R, Zhang M Z, Liu Y L, et al. AlphaFold, artificial intelligence (AI), and allostery[J]. The Journal of Physical Chemistry B, 2022, 126 (34): 6372-6383.

[192] Li Y J, Choi D, Chung J, et al. Competition-level code generation with AlphaCode[J]. Science, 2022, 378 (6624): 1092-1097.

[193] El-Sherif D M, Abouzid M, Elzarif M T, et al. Telehealth and artificial intelligence insights into healthcare during the COVID-19 pandemic[J]. Healthcare, 2022, 10 (2): 385.

[194] Assael Y, Sommerschield T, Shillingford B, et al. Restoring and attributing ancient texts using deep neural networks[J]. Nature, 2022, 603 (7900): 280-283.

[195] Anantrasirichai N, Bull D. Artificial intelligence in the creative industries: A review[J]. Artificial Intelligence Review, 2022, 55 (1): 589-656.

[196] Aydın Ö, Karaarslan E. OpenAI ChatGPT generated literature review: Digital twin in healthcare[M]//Aydın Ö. Emerging Computer Technologies 2. İzmir: İzmir Akademi Dernegi, 2022: 22-31.

[197] Rudolph J, Tan S, Tan S. ChatGPT: Bullshit spewer or the end of traditional assessments in higher education？[J]. Journal of Applied Learning and Teaching, 2023, 6 (1): 1-22.

[198] Holden Thorp H. ChatGPT is fun, but not an author[J]. Science, 2023, 379 (6630): 313.

[199] Lin H Y. Standing on the shoulders of AI giants[J]. Computer, 2023, 56 (1): 97-101.

[200] Xu Y J, Liu X, Cao X, et al. Artificial intelligence: a powerful paradigm for scientific research[J]. The Innovation, 2021, 2 (4): 100179.

[201] 张钹, 朱军, 苏航. 迈向第三代人工智能[J]. 中国科学: 信息科学, 2020, 50 (9): 1281-1302.

[202] Cox D R. The regression analysis of binary sequences[J]. Journal of the Royal Statistical Society Series B: Statistical Methodology, 1958, 20 (2): 215-232.

[203] Fix E, Hodges J L. Discriminatory analysis-nonparametric discrimination consistency properties[J]. International Statistical Review, 1951, 57: 238-247.

[204] Broomhead D S, Lowe D. Multivariable functional interpolation and adaptive networks[J]. Complex Systems, 1988, 2: 321-355.

[205] Creswell A, White T, Dumoulin V, et al. Generative adversarial networks: An overview[J]. IEEE Signal Processing Magazine, 2018, 35 (1): 53-65.

[206] Goodfellow I, Pouget-Abadie J, Mirza M, et al. Generative adversarial networks[J]. Communications of the ACM, 2020, 63 (11): 139-144.

[207] Bakrania A, Joshi N, Zhao X, et al. Artificial intelligence in liver cancers: Decoding the impact of machine learning models in clinical diagnosis of primary liver cancers and liver cancer metastases[J]. Pharmacological Research, 2023, 189: 106706.

[208] Mitchell T M. Machine Learning[M]. New York: McGraw-Hill, 1997.

[209] Vapnik V N. Statistical Learning Theory[M]. New York: Wiley, 1998.

[210] Jordan M I, Mitchell T M. Machine learning: Trends, perspectives, and prospects[J]. Science, 2015, 349 (6245): 255-260.

[211] 刘本永. 数字视听资料分析检验[M]. 北京: 科学出版社, 2022.

[212] Poggio T, Girosi F. Networks for approximation and learning[J]. Proceedings of the IEEE, 1990, 78 (9): 1481-1497.

[213] Allen D M. The relationship between variable selection and data agumentation and a method for prediction[J]. Technometrics, 1974, 16 (1): 125-127.

[214] Akaike H. A new look at the statistical model identification[J]. IEEE Transactions on Automatic Control, 1974, 19 (6): 716-723.

[215] Schwarz G. Estimating the dimension of a model[J]. The Annals of Statistics, 1978, 6 (2): 461-464.

[216] Hannan E J, Quinn B G. The determination of the order of an autoregression[J]. Journal of the Royal Statistical Society Series B: Statistical Methodology, 1979, 41 (2): 190-195.

[217] Wahba G. Spline Models for Observational Data[M]. Philadelphia: Society for Industrial and Applied Mathematics, 1990.

[218] Geman S, Bienenstock E, Doursat R. Neural networks and the bias/variance dilemma[J]. Neural Computation, 1992, 4 (1): 1-58.

[219] Stoica P, Selen Y. Model-order selection: A review of information criterion rules[J]. IEEE Signal Processing Magazine, 2004, 21 (4): 36-47.

[220] Sugiyama M, Ogawa H. Subspace information criterion for model selection[J]. Neural Computation, 2001, 13 (8): 1863-1889.

[221] Miyato T, Maeda S I, Koyama M, et al. Virtual adversarial training: A regularization method for supervised and semi-supervised learning[J]. IEEE Transactions on Pattern Analysis and Machine Intelligence, 2019, 41 (8): 1979-1993.

[222] Ding J, Tarokh V, Yang Y H. Model selection techniques: An overview[J]. IEEE Signal Processing Magazine, 2018, 35 (6): 16-34.

[223] Schaffer J. What not to multiply without necessity[J]. Australasian Journal of Philosophy, 2015, 93 (4): 644-664.

[224] Sra S, Nowozin S, Wright S J. Optimization for Machine Learning[M]. Cambridge: MIT Press, 2012.

[225] Bottou L, Curtis F E, Nocedal J. Optimization methods for large-scale machine learning[J]. SIAM Review, 2018, 60 (2): 223-311.

[226] Sun S L, Cao Z H, Zhu H, et al. A survey of optimization methods from a machine learning perspective[J]. IEEE Transactions on Cybernetics, 2020, 50 (8): 3668-3681.

[227] Novikoff A B. On convergence proofs for perceptrons[J]. Proceedings of the Symposium on Mathematical Theory of Automata, 1962, 12: 615-622.

[228] LeCun Y, Bottou L, Bengio Y, et al. Gradient-based learning applied to document recognition[J]. Proceedings of the IEEE, 1998, 86 (11): 2278-2324.

[229] Robbins H, Monro S. A stochastic approximation method[J]. The Annals of Mathematical Statistics, 1951, 22 (3): 400-407.

[230] Mnih V, Kavukcuoglu K, Silver D, et al. Human-level control through deep reinforcement learning[J]. Nature, 2015, 518 (7540): 529-533.

[231] Tieleman T, Hinton G. Divide the gradient by a running average of its recent magnitude[J]. COURSERA Neural Networks and Machine Learnning, 2012, 4 (2): 26-31.

[232] Gauss C F. Theory of the Motion of the Heavenly Bodies Moving about the Sun in Conic Sections[M]. New York: Dover, 1809.

[233] Wedderburn R W M. Quasi-likelihood functions, generalized linear models, and the Gauss—

Newton method[J]. Biometrika, 1974, 61 (3): 439-447.
[234] Wang Y. Gauss-Newton method[J]. Wiley Interdisciplinary Reviews: Computational Statistics, 2012, 4 (4): 415-420.
[235] Botev A, Ritter H, Barber D. Practical Gauss-Newton optimisation for deep learning [C]// Proceedings of the 34th International Conference on Machine Learning. New York: ACM, 2017, 70: 557-565.
[236] Chen C, Reiz S, Yu C D, et al. Fast approximation of the Gauss-Newton Hessian matrix for the multilayer perceptron[J]. SIAM Journal on Matrix Analysis and Applications, 2021, 42 (1): 165-184.
[237] Zamzam A S, Fu X, Sidiropoulos N D. Data-driven learning-based optimization for distribution system state estimation[J]. IEEE Transactions on Power Systems, 2019, 34 (6): 4796-4805.
[238] Mazyavkina N, Sviridov S, Ivanov S, et al. Reinforcement learning for combinatorial optimization: A survey[J]. Computers & Operations Research, 2021, 134: 105400.
[239] Halbouni A, Gunawan T S, Habaebi M H, et al. Machine learning and deep learning approaches for CyberSecurity: a review[J]. IEEE Access, 2022, 10: 19572-19585.
[240] Belson W A. Matching and prediction on the principle of biological classification[J]. Applied Statistics, 1959, 8 (2): 65-75.
[241] Safavian S R, Landgrebe D. A survey of decision tree classifier methodology[J]. IEEE Transactions on Systems, Man, and Cybernetics, 1991, 21 (3): 660-674.
[242] Ho Y C, Agrawala A K. On pattern classification algorithms introduction and survey[J]. Proceedings of the IEEE, 1968, 56 (12): 2101-2114.
[243] Toussaint G T. On a simple Minkowski metric classifier[J]. IEEE Transactions on Systems Science and Cybernetics, 1970, 6 (4): 360-362.
[244] Karhunen K. Über lineare methoden in der Wahrscheinlichkeitsrechnung[J]. Annals of Academic Science Fennicae, Series AI: Mathematics and Physics, 1946, 37: 3-79.
[245] Loève M. Fonctions Aléatoires du second ordre[M]//Lévy P. Processus Stochastiques et Movement Brownien. Paris: Hermann, 1948: 336-420.
[246] Fisher R A. The use of multiple measurements in taxonomic problems[J]. Annals of Eugenics, 1936, 7 (2): 179-188.
[247] Hartigan J A, Wong M A. Algorithm AS 136: A k-means clustering algorithm[J]. Journal of the Royal Statistical Society. Series C (Applied Statistics), 1979, 28 (1): 100-108.
[248] Kaelbling L P, Littman M L, Moore A W. Reinforcement learning: A survey[J]. Journal of Artificial Intelligence Research, 1996, 4: 237-285.
[249] Nguyen T T, Nguyen N D, Nahavandi S. Deep reinforcement learning for multiagent systems: A review of challenges, solutions, and applications[J]. IEEE Transactions on Cybernetics, 2020, 50 (9): 3826-3839.
[250] Board R, Pitt L. Semi-supervised learning[J]. Machine Learning, 1989, 4 (1): 41-65.
[251] van Engelen J E, Hoos H H. A survey on semi-supervised learning[J]. Machine Learning, 2020, 109 (2): 373-440.
[252] Juang B H, Katagiri S. Discriminative learning for minimum error classification (pattern

recognition)[J]. IEEE Transactions on Signal Processing, 1992, 40 (12): 3043-3054.
[253] Zheng Z D, Yang X D, Yu Z D, et al. Joint discriminative and generative learning for person re-identification[C]//2019 IEEE/CVF Conference on Computer Vision and Pattern Recognition (CVPR). New York: IEEE, 2019: 2133-2142.
[254] Cohn D, Atlas L, Ladner R. Improving generalization with active learning[J]. Machine Learning, 1994, 15 (2): 201-221.
[255] Cohn D A, Ghahramani Z, Jordan M I. Active learning with statistical models[J]. Journal of Artificial Intelligence Research, 1996, 4: 129-145.
[256] Gal Y, Islam R, Ghahramani Z. Deep Bayesian active learning with image data[C]//Proceedings of the 34th International Conference on Machine Learning-Volume 70. New York: ACM, 2017: 1183-1192.
[257] Ren P Z, Xiao Y, Chang X J, et al. A survey of deep active learning[J]. ACM Computing Surveys, 2021, 54 (9): 1-40.
[258] Bengio Y, Courville A, Vincent P. Representation learning: A review and new perspectives[J]. IEEE Transactions on Pattern Analysis and Machine Intelligence, 2013, 35 (8): 1798-1828.
[259] Rebuffi S A, Kolesnikov A, Sperl G, et al. iCaRL: Incremental classifier and representation learning[C]//2017 IEEE Conference on Computer Vision and Pattern Recognition (CVPR). New York: IEEE, 2017: 5533-5542.
[260] Zhang D K, Yin J, Zhu X Q, et al. Network representation learning: A survey[J]. IEEE Transactions on Big Data, 2020, 6 (1): 3-28.
[261] Kazemi S M, Goel R, Jain K, et al. Representation learning for dynamic graphs: A survey[J]. Journal of Machine Learning Research, 2020, 21 (70): 1-73.
[262] Merriman M. On the history of the method of least squares[J]. The Analyst, 1877, 4 (2): 33-36.
[263] Yule G U. On the theory of correlation[J]. Journal of the Royal Statistical Society, 1897, 60 (4): 812-854.
[264] Aitken A C. On least squares and linear combination of observations[J]. Proceedings of the Royal Society of Edinburgh, 1936, 55: 42-48.
[265] Levenberg K. A method for the solution of certain non-linear problems in least squares[J]. Quarterly of Applied Mathematics, 1944, 2 (2): 164-168.
[266] Cochrane D, Orcutt G H. Application of least squares regression to relationships containing auto-correlated error terms[J]. Journal of the American Statistical Association, 1949, 44 (245): 32-61.
[267] Plackett R L. A historical note on the method of least squares[J]. Biometrika, 1949, 36 (3/4): 458-460.
[268] Plackett R L. Some theorems in least squares[J]. Biometrika, 1950, 37 (1/2): 149-157.
[269] Marquardt D W. An algorithm for least-squares estimation of nonlinear parameters[J]. Journal of the Society for Industrial and Applied Mathematics, 1963, 11 (2): 431-441.
[270] Jennrich R I. Asymptotic properties of non-linear least squares estimators[J]. The Annals of Mathematical Statistics, 1969, 40 (2): 633-643.
[271] Sorenson H W. Least-squares estimation: from Gauss to Kalman[J]. IEEE Spectrum, 1970, 7 (7):

63-68.

[272] Cleveland W S. Robust locally weighted regression and smoothing scatterplots[J]. Journal of the American Statistical Association, 1979, 74 (368): 829-836.

[273] Golub G H, van Loan C F. An analysis of the total least squares problem[J]. SIAM Journal on Numerical Analysis, 1980, 17 (6): 883-893.

[274] Tarantola A, Valette B. Generalized nonlinear inverse problems solved using the least squares criterion[J]. Reviews of Geophysics, 1982, 20 (2): 219-232.

[275] Geladi P, Kowalski B R. Partial least-squares regression: A tutorial[J]. Analytica Chimica Acta, 1986, 185: 1-17.

[276] Chen S, Billings S A, Luo W. Orthogonal least squares methods and their application to non-linear system identification[J]. International Journal of Control, 1989, 50 (5): 1873-1896.

[277] Chen S, Cowan C N, Grant P M. Orthogonal least squares learning algorithm for radial basis function networks[J]. IEEE Transactions on Neural Networks, 1991, 2 (2): 302-309.

[278] Suykens J, Vandewalle J. Least squares support vector machine classifiers[J]. Neural Processing Letters, 1999, 9: 293-300.

[279] Suykens J A K, De Brabanter J, Lukas L, et al. Weighted least squares support vector machines: robustness and sparse approximation[J]. Neurocomputing, 2002, 48 (1/2/3/4): 85-105.

[280] Poggio T. On optimal nonlinear associative recall[J]. Biological Cybernetics, 1975, 19 (4): 201-209.

[281] Mao X D, Li Q, Xie H R, et al. On the effectiveness of least squares generative adversarial networks[J]. IEEE Transactions on Pattern Analysis and Machine Intelligence, 2019, 41 (12): 2947-2960.

[282] Sinaga M A, Stefanus L Y. Least square adversarial autoencoder[C]//2020 International Conference on Advanced Computer Science and Information Systems (ICACSIS). New York: IEEE, 2020: 33-40.

[283] Kohonen T, Ruohonen M. Representation of associated data by matrix operators[J]. IEEE Transactions on Computers, 1973, C-22 (7): 701-702.

[284] Moore E H. On the reciprocal of the general algebraic matrix[J]. Bulletin of American Mathematics Society, 1920, 2 (26): 394-395.

[285] Penrose R. A generalized inverse for matrices[J]. Mathematical Proceedings of the Cambridge Philosophical Society, 1955, 51 (3): 406-413.

[286] Albert A. Regression and The Moore-Penrose Pseudoinverse[M]. New York: Academic Press, 1972.

[287] Cooper L N, Liberman F, Oja E. A theory for the acquisition and loss of neuron specificity in visual cortex[J]. Biological Cybernetics, 1979, 33 (1): 9-28.

[288] Personnaz L, Guyon I, Dreyfus G. Collective computational properties of neural networks: New learning mechanisms[J]. Physical Review A, General Physics, 1986, 34 (5): 4217-4228.

[289] Michel A N, Farrell J A. Associative memories via artificial neural networks[J]. IEEE Control Systems Magazine, 1990, 10 (3): 6-17.

[290] Amari S I. Mathematical theory of neural learning[J]. New Generation Computing, 1991, 8 (4):

281-294.

[291] Gorodnichy D O, Reznik A M. Increasing attraction of pseudo-inverse autoassociative networks[J]. Neural Processing Letters, 1997, 5 (2): 51-55.

[292] Lee D L. Improvements of complex-valued hopfield associative memory by using generalized projection rules[J]. IEEE Transactions on Neural Networks, 2006, 17 (5): 1341-1347.

[293] Kobayashi M. Noise robust projection rule for hyperbolic hopfield neural networks[J]. IEEE Transactions on Neural Networks and Learning Systems, 2020, 31 (1): 352-356.

[294] Waterman D A. Generalization learning techniques for automating the learning of heuristics[J]. Artificial Intelligence, 1970, 1 (1/2): 121-170.

[295] Mitchell T M. Generalization as search[J]. Artificial Intelligence, 1982, 18 (2): 203-226.

[296] Werbos P J. Generalization of backpropagation with application to a recurrent gas market model[J]. Neural Networks, 1988, 1 (4): 339-356.

[297] Levin E, Tishby N, Solla S A. A statistical approach to learning and generalization in layered neural networks[J]. Proceedings of the IEEE, 1990, 78 (10): 1568-1574.

[298] Amari S I. Training error, generalization error and learning curves in neural learning[C]// Proceedings of the 2nd New Zealand Two-Stream International Conference on Artificial Neural Networks and Expert Systems. New York: ACM, 1995.

[299] Ivakhnenko A G. Polynomial theory of complex systems[J]. IEEE Transactions on Systems, Man, and Cybernetics, 1971, SMC-1 (4): 364-378.

[300] Wherry R J. Underprediction from overfitting: 45 years of shrinkage[J]. Personnel Psychology, 1975, 28 (1): 1-18.

[301] Dietterich T. Overfitting and undercomputing in machine learning[J]. ACM Computing Surveys, 1995, 27 (3): 326-327.

[302] Lawrence S, Giles C L, Tsoi A C. Lessons in neural network training: Overfitting may be harder than expected[C]//Proceedings of the Fourteenth National Conference on Artificial Intelligence and Ninth Conference on innovative Applications of Artificial Intelligence.New York: ACM, 1997: 540-545.

[303] Caruana R, Lawrence S, Giles L. Overfitting in neural nets: Backpropagation, conjugate gradient, and early stopping[C]//Proceedings of the 13th International Conference on Neural Information Processing Systems.New York: ACM, 2000: 381-387.

[304] Hawkins D M. The problem of overfitting[J]. Journal of Chemical Information and Computer Sciences, 2004, 44 (1): 1-12.

[305] Srivastava N, Hinton G, Krizhevsky A, et al. Dropout: A simple way to prevent neural networks from overfitting[J]. Journal of Machine Learning Research, 2014, 15: 1929-1958.

[306] Bartlett P L, Montanari A, Rakhlin A. Deep learning: A statistical viewpoint[J]. Acta Numerica, 2021, 30: 87-201.

[307] Hastie T, Montanari A, Rosset S, et al. Surprises in high-dimensional ridgeless least squares interpolation[J]. Annals of Statistics, 2022, 50 (2): 949-986.

[308] Särelä J, Vigário R. Overlearning in marginal distribution-based ICA: Analysis and solutions[J]. Journal of Machine Learning Research, 2003, 4: 1447-1469.

[309] Panchal G, Ganatra A, Shah P, et al. Determination of over-learning and over-fitting problem in back propagation neural network[J]. International Journal on Soft Computing, 2011, 2 (2): 40-51.

[310] Sjöberg J, Ljung L. Overtraining, regularization, and searching for minimum in neural networks[J]. IFAC Proceedings Volumes, 1992, 25 (14): 73-78.

[311] Tikhonov A N. Solution of incorrectly formulated problems and the regularization method[J]. Soviet Mathematic Doklady, 1963, 5: 1035-1038.

[312] Duffy D G. Green's Functions with Applications[M]. Boca Raton, FL, USA: CRC Press, 2015.

[313] Meixner J. Die greensche funktion des wellenmechanischen keplerproblems[J]. Mathematische Zeitschrift, 1933, 36 (1): 677-707.

[314] Hostler L, Pratt R H. Coulomb Green's function in closed form[J]. Physical Review Letters, 1963, 10 (11): 469-470.

[315] Hoerl A E, Kennard R W. Ridge regression: Biased estimation for nonorthogonal problems[J]. Technometrics, 1970, 12 (1): 55-67.

[316] Hoerl A E, Kennard R W. Ridge regression: Applications to nonorthogonal problems[J]. Technometrics, 1970, 12 (1): 69-82.

[317] Vinod H D. A survey of ridge regression and related techniques for improvements over ordinary least squares[J]. The Review of Economics and Statistics, 1978, 60 (1): 121-131.

[318] Golub G H, Heath M, Wahba G. Generalized cross-validation as a method for choosing a good ridge parameter[J]. Technometrics, 1979, 21 (2): 215-223.

[319] Saunders C, Gammerman A, Vovk V. Ridge regression learning algorithm in dual variables[C]// Proceedings of the Fifteenth International Conference on Machine Learning. New York: ACM, 1998: 515-521.

[320] Bishop C M. Training with noise is equivalent to Tikhonov regularization[J]. Neural Computation, 1995, 7 (1): 108-116.

[321] Sugiyama M, Ogawa H. Optimal design of regularization term and regularization parameter by subspace information criterion[J]. Neural Networks, 2002, 15 (3): 349-361.

[322] Huber P J. Robust estimation of a location parameter[J]. The Annals of Mathematical Statistics, 1964, 35 (1): 73-101.

[323] Ganapathiraju A, Hamaker J E, Picone J. Applications of support vector machines to speech recognition[J]. IEEE Transactions on Signal Processing, 2004, 52 (8): 2348-2355.

[324] Sánchez-Fernández M, de-Prado-Cumplido M, Arenas-Garcia J, et al. SVM multiregression for nonlinear channel estimation in multiple-input multiple-output systems[J]. IEEE Transactions on Signal Processing, 2004, 52 (8): 2298-2307.

[325] Liang Y C, Sun Y F. An improved method of support vector machine and its applications to financial time series forecasting[J]. Progress in Natural Science, 2003, 13 (9): 696-700.

[326] Wang L P. Support Vector Machines: Theory and Applications[M]. Berlin: Springer, 2005.

[327] Cervantes J, Garcia-Lamont F, Rodríguez-Mazahua L, et al. A comprehensive survey on support vector machine classification: Applications, challenges and trends[J]. Neurocomputing, 2020, 408: 189-215.

[328] Schölkopf B, Smola A J. Learning with Kernels: Support Vector Machines, Regularization, Optimization, and Beyond[M]. Cambridge: MIT Press, 2001.
[329] Honeine P, Richard C. Preimage problem in kernel-based machine learning[J]. IEEE Signal Processing Magazine, 2011, 28 (2): 77-88.
[330] Subrahmanya N, Shin Y C. Sparse multiple kernel learning for signal processing applications[J]. IEEE Transactions on Pattern Analysis and Machine Intelligence, 2010, 32 (5): 788-798.
[331] Chen B D, Zhao S L, Zhu P P, et al. Quantized kernel least mean square algorithm[J]. IEEE Transactions on Neural Networks and Learning Systems, 2012, 23 (1): 22-32.
[332] Okwuashi O, Ndehedehe C E. Deep support vector machine for hyperspectral image classification[J]. Pattern Recognition, 2020, 103: 107298.
[333] Suykens J A K, Lukas L, Van Dooren P, et al. Least squares support vector machine classifers: a large scale algorithm[J]. European Conference on Circuit Theory and Design, 1999: 839-842.
[334] Mohammadi M, Mousavi S H, Effati S. Generalized variant support vector machine[J]. IEEE Transactions on Systems, Man, and Cybernetics: Systems, 2021, 51 (5): 2798-2809.
[335] Niu D B, Wang C J, Tang P P, et al. An efficient algorithm for a class of large-scale support vector machines exploiting hidden sparsity[J]. IEEE Transactions on Signal Processing, 2022, 70: 5608-5623.
[336] Suykens J A K, Vandewalle J. Multiclass least squares support vector machines[C]//IJCNN'99. International Joint Conference on Neural Networks. Proceedings. New York: IEEE, 1999: 900-903.
[337] Van Den Burg G J J, Groenen P J F. GenSVM: A generalized multiclass support vector machine[J]. Journal of Machine Learning Research, 2016, 17: 1-42.
[338] Li S Y, Cai M Q, Mei L, et al. Multiclass weighted least squares twin bounded support vector machine for intelligent water leakage diagnosis[J]. IEEE Transactions on Instrumentation and Measurement, 2023, 72: 3514015.
[339] Wang D, Qiao H, Zhang B, et al. Online support vector machine based on convex hull vertices selection[J]. IEEE Transactions on Neural Networks and Learning Systems, 2013, 24 (4): 593-609.
[340] Wang J, Yang D W, Jiang W, et al. Semisupervised incremental support vector machine learning based on neighborhood kernel estimation[J]. IEEE Transactions on Systems, Man, and Cybernetics: Systems, 2017, 47 (10): 2677-2687.
[341] Xu J, Xu C, Zou B, et al. New incremental learning algorithm with support vector machines[J]. IEEE Transactions on Systems, Man, and Cybernetics: Systems, 2019, 49 (11): 2230-2241.
[342] Xu D J, Jiang M, Hu W Z, et al. An online prediction approach based on incremental support vector machine for dynamic multiobjective optimization[J]. IEEE Transactions on Evolutionary Computation, 2022, 26 (4): 690-703.
[343] Wintz P A. Transform picture coding[J]. Proceedings of the IEEE, 1972, 60 (7): 809-820.
[344] Gerbrands J J. On the relationships between SVD, KLT and PCA[J]. Pattern Recognition, 1981, 14 (1-6): 375-381.
[345] Hotelling H. Analysis of a complex of statistical variables into principal components[J]. Journal

of Educational Psychology, 1933, 24 (6): 417-441.
[346] Oja E. A simplified neuron model as a principal component analyzer[J]. Journal of Mathematical Biology, 1982, 15 (3): 267-273.
[347] Oja E, Karhunen J. On stochastic approximation of the eigenvectors and eigenvalues of the expectation of a random matrix[J]. Journal of Mathematical Analysis and Applications, 1985, 106 (1): 69-84.
[348] Sanger T D. Optimal unsupervised learning in a single-layer linear feedforward neural network[J]. Neural Networks, 1989, 2 (6): 459-473.
[349] Sanger T D. Two iterative algorithms for computing the singular value decomposition from input/output samples[C]//Proceedings of the 6th International Conference on Neural Information Processing Systems.New York: ACM, 1993: 144-151.
[350] Artac M, Jogan M, Leonardis A. Incremental PCA for on-line visual learning and recognition[C]//2002 International Conference on Pattern Recognition.New York: IEEE, 2002: 781-784.
[351] Weng J Y, Zhang Y L, Hwang W S. Candid covariance-free incremental principal component analysis[J]. IEEE Transactions on Pattern Analysis and Machine Intelligence, 2003, 25 (8): 1034-1040.
[352] Kim K I, Franz M O, Schölkopf B. Iterative kernel principal component analysis for image modeling[J]. IEEE Transactions on Pattern Analysis and Machine Intelligence, 2005, 27 (9): 1351-1366.
[353] Honeine P. Online kernel principal component analysis: A reduced-order model[J]. IEEE Transactions on Pattern Analysis and Machine Intelligence, 2012, 34 (9): 1814-1826.
[354] Ogawa H. Projection filter regularization of ill-conditioned problem[C]//Inverse Problems in Optics.Washington: SPIE, 1987, 808: 189-196.
[355] Oja E, Ogawa H. Parametric projection filter for image and signal restoration[J]. IEEE Transactions on Acoustics, Speech, and Signal Processing, 1986, 34 (6): 1643-1653.
[356] Ogawa H, Oja E. Optimally generalizing neural networks[C]//1991 IEEE International Joint Conference on Neural Networks. New York: IEEE, 1991: 2050-2055.
[357] Ogawa H. Neural network learning, generalization and over-learning[C]//Proceedings of International Conference on Intelligent Information Processing and Systems, Beijing, China, 1992, 2: 1-6.
[358] Vijayakumar S, Ogawa H. A functional analytic approach to incremental learning in optimally generalizing neural networks[C]//Proceedings of ICNN'95-International Conference on Neural Networks.New York: IEEE, 2002: 777-782.
[359] Sugiyama M, Ogawa H. Incremental active learning for optimal generalization[J]. Neural Computation, 2000, 12 (12): 2909-2940.
[360] Liu B Y. Adaptive training of a kernel-based nonlinear discriminator[J]. Pattern Recognition, 2005, 38 (12): 2419-2425.
[361] Liu B Y, Zhang J. An adaptively trained kernel-based nonlinear representor for handwritten digit classification[J]. Journal of Electronics (China), 2006, 23 (3): 379-383.

[362] Liu B Y, Zhang J, Chen X W. Adaptive training of a kernel-based representative and discriminative nonlinear classifier[C]//Liu D, Fei S, Hou Z, et al. International Symposium on Neural Networks. Berlin, Heidelberg: Springer, 2007: 381-390.

[363] Lamichhane B, Rebollo-Neira L. Projection and interpolation based techniques for structured and impulsive noise filtering[M]//Maeda T. New Signal Processing Research. New York: Nova Science Publishers, 2008: 1-32.

[364] Lee C, Eden M, Unser M. High-quality image resizing using oblique projection operators[J]. IEEE Transactions on Image Processing, 1998, 7 (5): 679-692.

[365] Hu Y, Zhang X, Zhu F, et al. Image recognition using iterative oblique projection[J]. Electronics Letters, 2005, 41 (20): 1109.

[366] Boyer R, Bouleux G. Oblique projections for direction-of-arrival estimation with prior knowledge[J]. IEEE Transactions on Signal Processing, 2008, 56 (4): 1374-1387.

[367] Varadarajan B, Khudanpur S, Tran T D. Stepwise optimal subspace pursuit for improving sparse recovery[J]. IEEE Signal Processing Letters, 2011, 18 (1): 27-30.

[368] Zuo W L, Xin J M, Liu W Y, et al. Localization of near-field sources based on linear prediction and oblique projection operator[J]. IEEE Transactions on Signal Processing, 2019, 67 (2): 415-430.

[369] Berger P, Gröchenig K, Matz G. Sampling and reconstruction in distinct subspaces using oblique projections[J]. Journal of Fourier Analysis and Applications, 2019, 25 (3): 1080-1112.

[370] Zhu K, Yu C P, Wan Y M. Recursive least squares identification with variable-direction forgetting via oblique projection decomposition[J]. IEEE/CAA Journal of Automatica Sinica, 2022, 9 (3): 547-555.

[371] Otto S E, Padovan A, Rowley C W. Optimizing oblique projections for nonlinear systems using trajectories[J]. SIAM Journal on Scientific Computing, 2022, 44 (3): A1681-A1702.

[372] Aronszajn N, Glazman I. Theory of Linear Operators in Hilbert Space[M]. New York: Frederic Ungar Publishing Company, 1963.

[373] Rynne B P, Youngson M A. Linear Functional Analysis[M]. London: Springer, 2000.

[374] Siddiqi A H. Applied Functional Analysis: Numerical Methods, Wavelet Methods, and Image Processing[M]. New York: Marcel Dekker, 2004.

[375] 吴应江. 基于多核学习的疑似 AD 脑 MRI 影像分类算法[D]. 贵阳: 贵州大学, 2017.

[376] Aronszajn N. Theory of reproducing kernels[J]. Transactions of the American Mathematical Society, 1950, 68 (3): 337-404.

[377] Saitoh S, Sawano Y. Theory of Reproducing Kernels and Applications[M]. Singapore: Springer Singapore, 2016.

[378] Schatten R. Norm Ideals of Completely Continuous Operators[M]. Berlin: Springer, 1960.

[379] Duffin R J, Schaeffer A C. A class of nonharmonic Fourier series[J]. Transactions of the American Mathematical Society, 1952, 72 (2): 341-366.

[380] Mallat S. A Wavelet Tour of Signal Processing[M]. San Diego: Academic Press, 1996.

[381] Gao J B, Harris C J, Gunn S R. On a class of support vector kernels based on frames in function Hilbert spaces[J]. Neural Computation, 2001, 13 (9): 1975-1994.

[382] Rakotomamonjy A, Canu S. Frames, reproducing kernels, regularization and learning[J]. Journal of Machine Learning Research, 2005, 6: 1485-1515.

[383] Tseng Y Y. The Characteristic Value Problem of Hermitian Functional Operators in a Non-Hilbert Space[D]. Chicago: University of Chicago, 1933.

[384] Beutler F J. The operator theory of the pseudo-inverse I. Bounded operators[J]. Journal of Mathematical Analysis and Applications, 1965, 10 (3): 451-470.

[385] Beutler F J. The operator theory of the pseudo-inverse II. Unbounded operators with arbitrary range[J]. Journal of Mathematical Analysis and Applications, 1965, 10 (3): 471-493.

[386] Groetsch C W. Generalized Inverses of Linear Operators: Representation and Approximation[M]. New York: Marcel Dekker, 1977.

[387] Christensen O. Frames and pseudo-inverses[J]. Journal of Mathematical Analysis and Applications, 1995, 195 (2): 401-414.

[388] Desore C A, Whalen B H. A note on pseudoinverses[J]. Journal of Society for Industrial and Applied Mathematics, 1963, 11 (2): 442-447.

[389] Pyt'ev Y P. Pseudoinverse operators: Properties and applications[J]. Mathematics of the USSR-Sbornik, 1983, 46 (1): 17-50.

[390] Nashed M Z, Votruba G F. A unified approach to generalized inverses of linear operators: I. Algebraic, topological, and projectional properties[J]. Bulletin of the American Mathematical Society, 1974, 80 (5): 825-830.

[391] Christensen O. Frames and the projection method[J]. Applied and Computational Harmonic Analysis, 1993, 1 (1): 50-53.

[392] Coates A, Ng A, Lee H. An analysis of single-layer networks in unsupervised feature learning[J]. Proceedings of Machine Learning Research, 2011, 15: 215-223.

[393] Ogawa H, Oja E. Projection filter, Wiener filter, and Karhunen-Loève subspaces in digital image restoration[J]. Journal of Mathematical Analysis and Applications, 1986, 114 (1): 37-51.

[394] Liu B Y, Ogawa H. An equivalent form of S-L projection learning[J]. Journal of Electronic Science and Technology of China, 2003, 1 (1): 6-11.

[395] Zhang J, Liu B Y, and Tan H. A kernel-based nonlinear representor for eigenface classification[J]. Journal of Electronic Science and Technology of China, 2004, 2 (2): 19-22.

[396] Liu B Y, Zhang J. Face recognition applying a kernel-based representative and discriminative nonlinear classifier to eigenspectra[C]//Proceedings of 2005 International Conference on Communications, Circuits and Systems. New York: IEEE, 2005: 964-968.

[397] Vijayakumar S, Sugiyama M, Ogawa H. Training data selection for optimal generalization with noise variance reduction in neural networks[C]//Marinaro M, Tagliaferri R. Neural Nets WIRN VIETRI-98. London: Springer, 1999: 153-166.

[398] Koide Y, Yamashita Y, Ogawa H. A unified theory of the family of projection filters for signal and image estimation[J]. Systems and Computers in Japan, 1995, 26 (4): 95-105.

[399] Yamashita Y, Ogawa H. Image restoration by averaged projection filter[J]. Systems and Computers in Japan, 1992, 23 (1): 79-88.

[400] Ogawa H. Image resoration by partial projection filter[J]. IEICE Transactions on Fundamentals,

1988, 71 (2): 519-526.
[401] Ogawa H, Hara S. Properties of partial projection filter[J]. IEICE Transactions on Fundamentals, 1988, 71 (2): 527-533.
[402] Ogawa H, Oja E, Lampinen J. Projection filters for image and signal restoration[C]//IEEE 1989 International Conference on Systems Engineering. New York: IEEE, 1989: 93-97.
[403] Ogawa H, Yamasaki K. Generalization and over-learning of neural networks[R]. Technical Report of IEICE, 1992, NC91-75 (1): 77-84.
[404] Hirabayashi A, Ogawa H. A class of learning for optimal generalization[C]//IJCNN'99. International Joint Conference on Neural Networks. Proceedings. New York: IEEE, 1999: 1815-1819.
[405] Hirabayashi A, Ogawa H. A family of projection learnings[J]. Systems and Computers in Japan, 2001, 32 (5): 21-35.
[406] Youla D. Generalized image restoration by the method of alternating orthogonal projections[J]. IEEE Transactions on Circuits and Systems, 1978, 25 (9): 694-702.
[407] Douglas R G. On majorization, factorization, and range inclusion of operators on Hilbert space[J]. Proceedings of the American Mathematical Society, 1966, 17 (2): 413.
[408] Ogawa H. Operator equations related to the restoration problems[J]. IEICE Technical Report, 1986, PRU86-60: 9-16.
[409] Xu Q X, Sheng L J, Gu Y Y. The solutions to some operator equations[J]. Linear Algebra and Its Applications, 2008, 429 (8/9): 1997-2024.
[410] Liu B Y, Zhang J. Partial oblique projection learning for optimal generalization[J]. Journal of Electronic Science and Technology of China, 2004, 2 (1): 63-68.
[411] de Kruif B J, de Vries T J A. Pruning error minimization in least squares support vector machines[J]. IEEE Transactions on Neural Networks, 2003, 14 (3): 696-702.
[412] Hoegaerts L, Suykens J A K, Vandewalle J, et al. A comparison of pruning algorithms for sparse least squares support vector machines[C]//Pal N R, Kasabov N, Mudi R K, et al. International Conference on Neural Information Processing. Berlin: Springer, 2004: 1247-1253.
[413] Wang H F, Hu D J. Comparison of SVM and LS-SVM for regression[C]//2005 International Conference on Neural Networks and Brain. New York: IEEE, 2005: 279-283.
[414] Yang J, Bouzerdoum A, Phung S L. A training algorithm for sparse LS-SVM using Compressive Sampling[C]//2010 IEEE International Conference on Acoustics, Speech and Signal Processing.New York: IEEE, 2010: 2054-2057.
[415] Afriat S N. Orthogonal and oblique projectors and the characteristics of pairs of vector spaces[J]. Mathematical Proceedings of the Cambridge Philosophical Society, 1957, 53 (4): 800-816.
[416] Kayalar S, Weinert H L. Oblique projections: Formulas, algorithms, and error bounds[J]. Mathematics of Control, Signals and Systems, 1989, 2 (1): 33-45.
[417] Behrens R T, Scharf L L. Signal processing applications of oblique projection operators[J]. IEEE Transactions on Signal Processing, 1994, 42 (6): 1413-1424.
[418] Park D C, El-Sharkawi M A, Marks R J. An adaptively trained neural network[J]. IEEE

Transactions on Neural Networks, 1991, 2 (3): 334-345.
[419] Fu L, Hsu H H, Principe J C. Incremental backpropagation learning networks[J]. IEEE Transactions on Neural Networks, 1996, 7 (3): 757-761.
[420] Giraud-Carrier C. A note on the utility of incremental learning[J]. AI Communications, 2000, 13 (4): 215-223.
[421] Losing V, Hammer B, Wersing H. Incremental on-line learning: A review and comparison of state of the art algorithms[J]. Neurocomputing, 2018, 275: 1261-1274.
[422] Sugiyama M, Ogawa H. Incremental projection learning for optimal generalization[J]. Neural Networks, 2001, 14 (1): 53-66.
[423] Liu B Y, Zhang J. Incremental POP learning[J]. Journal of Electronic Science and Technology of China, 2004, 2 (4): 29-36.
[424] Rao C R, Mitra S K. Generalized Inverse of Matrices and Its Applications[M]. New York: Wiley, 1971.
[425] Nilsson N J. Learning Machines: Foundations of Trainable Pattern-Classifying Systems[M]. New York: McGraw-Hill, 1965.
[426] Liu B Y. Kernel-based nonlinear discriminator with closed-form solution[C]//International Conference on Neural Networks and Signal Processing. New York: IEEE, 2003: 41-44.
[427] Mika S, Rätsch G, Weston J, et al. Fisher discriminant analysis with kernels[C]//Neural Networks for Signal Processing IX: Proceedings of the 1999 IEEE Signal Processing Society Workshop. New York: IEEE, 1999: 41-48.
[428] Barker M, Rayens W. Partial least squares for discrimination[J]. Journal of Chemometrics, 2003, 17 (3): 166-173.
[429] Brereton R G, Lloyd G R. Partial least squares discriminant analysis: Taking the magic away[J]. Journal of Chemometrics, 2014, 28 (4): 213-225.
[430] Dufrenois F, Noyer J C. Formulating robust linear regression estimation as a one-class LDA criterion: discriminative hat matrix[J]. IEEE Transactions on Neural Networks and Learning Systems, 2013, 24 (2): 262-273.
[431] Dufrenois F, Noyer J C. One class proximal support vector machines[J]. Pattern Recognition, 2016, 52: 96-112.
[432] Dufrenois F, Hamad D. Sparse and online null proximal discriminant analysis for one class learning in large-scale datasets[C]//2019 International Joint Conference on Neural Networks (IJCNN). New York: IEEE, 2019: 1-8.
[433] Liu B Y, Zhang J. Oblique projection realization of a kernel-based nonlinear discriminator[J]. Journal of Electronics (China), 2006, 23 (1): 94-98.
[434] Albert A. Conditions for positive and nonnegative definiteness in terms of pseudoinverses[J]. SIAM Journal on Applied Mathematics, 1969, 17 (2): 434-440.
[435] 刘本永. 基于投影逼近的分类器设计[C]//2009 年全国模式识别学术会议暨首届中日韩模式识别学术研讨会论文集. 北京: 中国自动化学会, 2009: 35-39.
[436] Liu B Y. Kernel discrimination via oblique projection[C]//2011 International Conference on Image Analysis and Signal Processing. New York: IEEE, 2011: 707-711.

[437] 刘本永．斜投影核鉴别器的增量学习：理论及算法[EB/OL]．(2012-10-29)[2024-05-01]. https://www.paper.eda.cn/releasepaper/content/201210-288.
[438] 刘本永．斜投影核鉴别器的增量学习：证明及示例[EB/OL]．(2013-11-14)[2024-05-01]. https://www.paper.edu.cn/releasepaper/content/201301-584.
[439] Gonzalez R C, Woods R E. Digital Image Processing, Second Edition[M]. Beijing: Publishing House of Electronics Industry, 2002.
[440] Pal N R, Pal S K. A review on image segmentation techniques[J]. Pattern Recognition, 1993, 26 (9): 1277-1294.
[441] Otsu N. A threshold selection method from gray-level histograms[J]. IEEE Transactions on Systems, Man, and Cybernetics, 1979, 9 (1): 62-66.
[442] Vincent L, Soille P. Watersheds in digital spaces: An efficient algorithm based on immersion simulations[J]. IEEE Transactions on Pattern Analysis and Machine Intelligence, 1991, 13 (6): 583-598.
[443] Liu B Y, Wu W Y, Chen X W. Kernel fitting for image segmentation[C]//2008 International Conference on Machine Learning and Cybernetics. New York: IEEE, 2008: 2914-2917.
[444] Kraaijveld M A. A Parzen classifier with an improved robustness against deviations between training and test data[J]. Pattern Recognition Letters, 1996, 17 (7): 679-689.
[445] Schonhoff T, Giordano A A. Detection and Estimation: Theory and Its Applications[M]. Beijing: Publishing House of Electronics Industry, 2007.
[446] Park S C, Park M K, Kang M G. Super-resolution image reconstruction: A technical overview[J]. IEEE Signal Processing Magazine, 2003, 20 (3): 21-36.
[447] 胡业刚，张毅，刘本永．小图像放大：算法与评价[J]．贵州大学学报（自然科学版），2010, 27 (2): 78-82.
[448] 崔静，刘本永．基于分组 SVR 和 KNR 的单帧图像超分辨[J]．计算机工程与应用，2012, 48 (23): 185-190.
[449] Harris C, Stephens M. A combined corner and edge detector[C]//Taylor C J. Proceedings of 4th Alvey Vision Conference, Manchester: Alvety Vision Club, 1988, 1: 147-151.
[450] Lowe D G. Distinctive image features from scale-invariant keypoints[J]. International Journal of Computer Vision, 2004, 60 (2): 91-110.
[451] Liu B Y, Zhang J, Liao X. Projection kernel regression for image registration and fusion in video-based criminal investigation[C]//2011 International Conference on Multimedia and Signal Processing.New York: IEEE, 2011: 348-352.
[452] Li Q, Zheng J S, Tsai A, et al. Robust endpoint detection and energy normalization for real-time speech and speaker recognition[J]. IEEE Transactions on Speech and Audio Processing, 2002, 10 (3): 146-157.
[453] Ephraim Y, Malah D. Speech enhancement using a minimum-mean square error short-time spectral amplitude estimator[J]. IEEE Transactions on Acoustics, Speech, and Signal Processing, 1984, 32 (6): 1109-1121.
[454] Stadtschnitzer M, Van Pham T, Chien T T. Reliable voice activity detection algorithms under adverse environments[C]//2008 Second International Conference on Communications and

Electronics.New York: IEEE, 2008: 218-223.
[455] Liu B Y, Zhang J, Liao X. Kernel fitting for speech detection and enhancement[C]//IEEE 10th International Conference on Signal Processing Proceedings. New York: IEEE, 2010: 534-537.
[456] Boll S. Suppression of acoustic noise in speech using spectral subtraction[J]. IEEE Transactions on Acoustics, Speech, and Signal Processing, 1979, 27 (2): 113-120.
[457] Jain A K, Duin R P W, Mao J C. Statistical pattern recognition: A review[J]. IEEE Transactions on Pattern Analysis and Machine Intelligence, 2000, 22 (1): 4-37.
[458] Chellappa R, Wilson C L, Sirohey S. Human and machine recognition of faces: A survey[J]. Proceedings of the IEEE, 1995, 83 (5): 705-741.
[459] Adjabi I, Ouahabi A, Benzaoui A, et al. Past, present, and future of face recognition: A review[J]. Electronics, 2020, 9 (8): 1188.
[460] Wang M, Deng W H. Deep face recognition: A survey[J]. Neurocomputing, 2021, 429: 215-244.
[461] Samaria F S, Harter A C. Parameterisation of a stochastic model for human face identification[C]//Proceedings of 1994 IEEE Workshop on Applications of Computer Vision.New York: IEEE, 1994: 138-142.
[462] Belhumeur P N, Hespanha J P, Kriegman D J. Eigenfaces vs. Fisherfaces: Recognition using class specific linear projection[J]. IEEE Transactions on Pattern Analysis and Machine Intelligence, 1997, 19 (7): 711-720.
[463] Liu B Y, Zhang J. Eigenspectra versus eigenfaces: classification with a kernel-based nonlinear representor[M]//Wang L P, Chen K, Ong Y S. Lecture Notes in Computer Science. Berlin: Springer, 2005: 660-663.
[464] Lu J W, Plataniotis K N, Venetsanopoulos A N. Face recognition using LDA-based algorithms[J]. IEEE Transactions on Neural Networks, 2003, 14 (1): 195-200.
[465] Tian Y, Tan T N, Wang Y H, et al. Do singular values contain adequate information for face recognition? [J]. Pattern Recognition, 2003, 36 (3): 649-655.
[466] Liu Q S, Lu H Q, Ma S D. A non-parameter Bayesian classifier for face recognition[J]. Journal of Electronics, 2003, 20 (5): 362-370.
[467] Chien J T, Wu C C. Discriminant waveletfaces and nearest feature classifiers for face recognition[J]. IEEE Transactions on Pattern Analysis and Machine Intelligence, 2002, 24 (12): 1644-1649.
[468] Liu X M, Chen T, Kumar B V K V. Face authentication for multiple subjects using eigenflow[J]. Pattern Recognition, 2003, 36 (2): 313-328.
[469] Atal B S. Automatic recognition of speakers from their voices[J]. Proceedings of the IEEE, 1976, 64 (4): 460-475.
[470] Reynolds D A, Rose R C. Robust text-independent speaker identification using Gaussian mixture speaker models[J]. IEEE Transactions on Speech and Audio Processing, 1995, 3 (1): 72-83.
[471] Chowdhury A, Ross A. Fusing MFCC and LPC features using 1D triplet CNN for speaker recognition in severely degraded audio signals[J]. IEEE Transactions on Information Forensics

and Security, 2019, 15: 1616-1629.
[472] Davis S. Order dependence in templates for monosyllabic word identification[C]//ICASSP '79. IEEE International Conference on Acoustics, Speech, and Signal Processing.New York: IEEE, 2003: 570-573.
[473] Kinnunen T, Karpov E, Franti P. Real-time speaker identification and verification[J]. IEEE Transactions on Audio, Speech, and Language Processing, 2006, 14 (1): 277-288.
[474] Bhanu B. Automatic target recognition: State of the art survey[J]. IEEE Transactions on Aerospace and Electronic Systems, 1986, AES-22 (4): 364-379.
[475] Roth M W. Survey of neural network technology for automatic target recognition[J]. IEEE Transactions on Neural Networks, 1990, 1 (1): 28-43.
[476] Rogers S K, Colombi J M, Martin C E, et al. Neural networks for automatic target recognition[J]. Neural Networks, 1995, 8 (7/8): 1153-1184.
[477] Nasrabadi N M. DeepTarget: An automatic target recognition using deep convolutional neural networks[J]. IEEE Transactions on Aerospace and Electronic Systems, 2019, 55 (6): 2687-2697.
[478] Bhanu B, Jones T L. Image understanding research for automatic target recognition[J]. IEEE Aerospace and Electronic Systems Magazine, 1993, 8 (10): 15-23.
[479] 周德全. 基于一维距离像的雷达目标识别研究[D]. 南京: 南京理工大学, 1998.
[480] 刘本永. 子空间法雷达目标一维像识别研究[D]. 成都: 电子科技大学, 1999.
[481] Jacobs S P, O'Sullivan J A. Automatic target recognition using sequences of high resolution radar range-profiles[J]. IEEE Transactions on Aerospace and Electronic Systems, 2000, 36 (2): 364-381.
[482] Zhao Q, Principe J C. Support vector machines for SAR automatic target recognition[J]. IEEE Transactions on Aerospace and Electronic Systems, 2001, 37 (2): 643-654.
[483] Martorella M, Giusti E, Demi L, et al. Target recognition by means of polarimetric ISAR images[J]. IEEE Transactions on Aerospace and Electronic Systems, 2011, 47 (1): 225-239.
[484] Anagnostopoulos G C. SVM-based target recognition from synthetic aperture radar images using target region outline descriptors[J]. Nonlinear Analysis: Theory, Methods & Applications, 2009, 71 (12): e2934-e2939.
[485] El-Darymli K, Gill E W, McGuire P, et al. Automatic target recognition in synthetic aperture radar imagery: A state-of-the-art review[J]. IEEE Access, 2016, 4: 6014-6058.
[486] Kechagias-Stamatis O, Aouf N. Automatic target recognition on synthetic aperture radar imagery: a survey[J]. IEEE Aerospace and Electronic Systems Magazine, 2021, 36 (3): 56-81.
[487] Turaga P, Chellappa R, Subrahmanian V S, et al. Machine recognition of human activities: A survey[J]. IEEE Transactions on Circuits and Systems for Video Technology, 2008, 18 (11): 1473-1488.
[488] Aggarwal J K, Ryoo M S. Human activity analysis: A review[J]. ACM Computing Surveys, 2011, 43 (3): 1-43.
[489] Ke S R, Thuc H, Lee Y J, et al. A review on video-based human activity recognition[J]. Computers, 2013, 2 (2): 88-131.
[490] Wang H, Kläser A, Schmid C, et al. Dense trajectories and motion boundary descriptors for

action recognition[J]. International Journal of Computer Vision, 2013, 103 (1): 60-79.

[491] Dalal N, Triggs B. Histograms of oriented gradients for human detection[C]//2005 IEEE Computer Society Conference on Computer Vision and Pattern Recognition (CVPR'05). New York: IEEE, 2005: 886-893.

[492] Dai Y, Shibata Y, Hashimoto K, et al. Facial expression recognition of person without language ability based on the optical flow histogram[C]//2000 5th International Conference on Signal Processing Proceedings.New York: IEEE, 2000: 1209-1212.

[493] Wang H, Schmid C. Action recognition with improved trajectories[C]//2013 IEEE International Conference on Computer Vision. New York: IEEE, 2013: 3551-3558.

[494] Sydorov V, Sakurada M, Lampert C H. Deep fisher kernels: End to end learning of the fisher kernel GMM parameters[C]//2014 IEEE Conference on Computer Vision and Pattern Recognition. New York: IEEE, 2014: 1402-1409.

[495] Gudovskiy D, Hodgkinson A, Yamaguchi T, et al. Deep active learning for biased datasets via fisher kernel self-supervision[C]//2020 IEEE/CVF Conference on Computer Vision and Pattern Recognition (CVPR). New York: IEEE, 2020: 9038-9046.

[496] Mairal J, Koniusz P, Harchaoui Z, et al. Convolutional kernel networks[C]//Proceedings of the 27th International Conference on Neural Information Processing Systems-Volume 2.New York: ACM, 2014: 2627-2635.

[497] Sahbi H. Kernel-based graph convolutional networks[C]//2020 25th International Conference on Pattern Recognition (ICPR). New York: IEEE, 2021: 4887-4894.

[498] Peng X J, Zou C Q, Qiao Y, et al. Action recognition with stacked fisher vectors[C]//European Conference on Computer Vision. Cham: Springer, 2014: 581-595.

[499] 熊壮. 视频行为分析中的深度核非线性网络[D]. 贵阳: 贵州大学, 2018.

[500] Jaakkola T S, Haussler D. Exploiting generative models in discriminative classifiers[C]// Proceedings of the 1998 conference on Advances in neural information processing systems II. Cambridge: MIT Press, 1999: 487-493.

[501] Perronnin F, Sánchez J, Mensink T. Improving the fisher kernel for large-scale image classification[C]//European Conference on Computer Vision. Berlin: Springer, 2010: 143-156.

[502] Nilsback M E, Zisserman A. A visual vocabulary for flower classification[C]//2006 IEEE Computer Society Conference on Computer Vision and Pattern Recognition (CVPR'06). New York: IEEE, 2006: 1447-1454.

[503] Gorelick L, Blank M, Shechtman E, et al. Actions as space-time shapes[J]. IEEE Transactions on Pattern Analysis and Machine Intelligence, 2007, 29 (12): 2247-2253.